BIOLINGUISTICS AND PHILOSOPHY: INSIGHTS AND OBSTACLES

ELLIOT MURPHY

BIOLINGUISTICS AND PHILOSOPHY

PUBLISHED BY LULU, 2012
COPYRIGHT © ELLIOT MURPHY, 2012. ALL RIGHTS RESERVED
ELLIOT MURPHY HAS ASSERTED HIS RIGHT UNDER THE COPYRIGHT, DESIGNS AND
PATENTS ACT 1988 TO BE IDENTIFIED AS THE AUTHOR OF THIS WORK
FIRST PUBLISHED IN LONDON, GREAT BRITAIN BY
LULU, 2012

WWW.LULU.COM
ISBN 978-1-291-18677-2

ELLIOT MURPHY

CONTENTS

Preface	4
1. 'I' BEFORE 'E'	6
2. LAWS OF FORM	44
3. SCOPES AND LIMITS	77
4. THE MENTAL AND THE NATURAL	111
5. THE HOLY TRINITY: REFERENCE, REPRESENTATIONALISM, EXTERNALISM	191
Index	263

PREFACE

We are animals sick with language. And how sometimes we long for a cure. But just shutting up won't do it. You can't just wish your way into animality. So it is then, as a matter of consolation, that we watch the animal channels and marvel at a world untamed by language. The animals get us to hear a voice of pure silence. Nostalgia for the fish life ... We record whales singing their whale songs capable of transmitting messages to other whales thousands of kilometers away, but in truth, brandishing our microphones, we only aspire to one thing – that those whales would sing us a song.

GÉRARD WAJCMAN[1]

The major division in the scientific world is between the natural and cognitive sciences. The former generally deals with what is outside the head, the latter with what is inside. The aspiration of many in both domains is to breach the gap between them, to the extent that the 'mental' world is explained to the same degree of empirical precision achieved by those who study the 'external' world. One might summarise that cognitive science and its branches, such as linguistics, use the mind to study itself. Concerns immediately arise over this division: is the study of the mind concerned with different 'physical' 'stuff' than the examination of the motions of molecules involved in the Haber process? Is there a material and naturalistic distinction between 'mind' and 'body,' or is it purely a conceptual one? These and other questions occupied a great deal of Enlightenment and Romantic thought. Such 'natural philosophers' as Descartes, Galileo, Newton, Hobbes, Locke, Leibniz, Bošković, Hume and Priestley have left us with the results of a great deal of empirical and a priori investigation into the nature of mechanical causation, celestial motion, psychic identity and unconscious principles of mind.

This study attempts to show how biolinguistics is the most fruitful and fascinating point of convergence between the two traditional fields of 'science' and 'philosophy.' If Anton Chekhov was right in saying that 'Man will become better when you show him what he is like,' then exploring where language came from and whether it influences the rest of 'thought' are likely to be domains of the highest interest, capable of solving many traditional philosophical problems.[2] Aside from Hinzen's *Mind Design and Minimal Syntax*, Boeckx's *Linguistic Minimalism* and a handful of papers mostly found in the journal *Biolinguistics*, studies of what might be called the 'philosophy of biolinguistics' have been few and short. Exploring the insights of recent theoretical linguists and neurobiologists can shed some much needed light on the problems posed by analytical philosophy. These concerns will take up the majority of the following pages, with a focus both on current research into evolutionary questions and on contemporary philosophy of language taking up the rest. The first two chapters will be broadly linguistic in focus, the next two being philosophical, and the final chapter a mixture of both. Because of the rich history of thought surrounding the study of language, I will typically allow quotations to speak directly to the reader, and I've tried not to blur certain passages with paraphrases or unnecessary commentary.

Along with modern philosophers, the 'natural philosophers' of the Enlightenment also studied questions of language and mind. The older these philosophers are, the harder it is to resist quoting them. And since the older philosophers typically put matters much more neatly and coherently then today's, I will liberally quote those seventeenth-century thinkers who were perceptive enough to fight their intuitions about how the world works and what it's made of (or their 'folk science') to come to conclusions which contemporary thinkers in biolinguistics wholeheartedly endorse. As André Gide wryly put the matter in 1891: 'Everything has been said before, but since nobody listens we have to keep going back and beginning it all over again.'[3]

REFERENCES

---

[1] Gérard Wajcman, 'The Animals that Treat Us Badly,' *Lacanian Ink*, no. 33: 131.
[2] Cited in Ben Schrank, 'Lunch with the FT: Steven Pinker,' *Financial Times*, 22 September 2002.
[3] Cited in Galen Strawson, *Real Materialism and Other Essays* (Oxford University Press, 2008), p. 4.

1

'I' BEFORE 'E'

For in spite of language, in spite of intelligence and intuition and sympathy, one can never really communicate anything to anybody.
ALDOUS HUXLEY[1]

Upon meeting his friend and student, Elizabeth Anscombe, in a corridor at Cambridge, Wittgenstein asked her: 'Tell me, why do people always say it was natural for man to assume that the sun went round the Earth rather than that the Earth was rotating?' Anscombe replied: 'Well, obviously because it just looks as though the sun is going round the Earth.' Wittgenstein responded: 'Well, what would it have looked like if it had looked as though the Earth was rotating?' If Anscombe found herself unable to reply, she had only her constrained and adapted wits to blame. The most important issue we can raise concerning why we think and experience the things we do is over how, in fact, the mind (what Darwin called the 'Citadel itself') comes to interpret the world to begin with. This genetically determined initial state of the body will shape how we see the sun 'rise' or a planes 'fly,' and other events in the external world which ultimately take shape through our general understanding of how to deal with experience.

Knowledge acquisition is, above all other traits, characteristic of humans. In his study of the Scottish intellectual tradition, George Davie recognised its central theme to be an understanding of the importance of 'natural beliefs or principles of common sense, such as the belief in an independent external world, the belief in causality, the belief in ideal standards, and the belief in the self of conscience as separate from the rest of one.'[2] Each of these concepts is rooted in our intuitive 'folk science' or 'common wisdom' about how the world works, which grows out of our intrinsic natural design, determining the range and limits of 'knowledge.' The study of such 'principles of common sense' has been dubbed 'Plato's problem' by Noam Chomsky – namely, why do we know so much with so little? This question reaches into of philosophy, neuroscience, linguistics, psychology, broader domains of biology and cognitive science, and many other inquiries of nature euphemistically called 'above the neck.' The problem was outlined best by Bertrand Russell, who asked 'How comes it that human beings, whose contacts with the world are brief and personal and limited, are nevertheless able to know so much as they do know?'[3]

To briefly spell out our terminology, it seems that the word 'subjective' appropriately means different things to different people, so I'll avoid it and instead stick to more specific terms in describing the internal structure and functions of the brain. By the 'mind' or its 'mental' faculties, I am the concerned with the abstract categorisation of the properties of the brain which, in the last half century, we have gradually begun to understand. Ambiguous definitions aside, one often feels like Eddington: 'I shall have to explain that by 'mind' I do not here exactly mean mind and by 'stuff' I do not at all mean stuff.'[4]

A short account of the problems surrounding language acquisition is also necessary before explaining the significance of the 'bio' in 'biolinguistics.' As is widely understood, the left hemisphere (LH) processes phonology, morphology, lexical information and syntax. Damage to it causes speech devoid of meaning and comprehension problems. The right hemisphere (RH) processes intonation, non-literal aspects of language (jokes, etc.) and emotion, with damage to it causing inference problems and impaired emotional interpretation. The famous Wernicke's area receives input for auditory and visual cortexes, and is the locus of semantic processing. Connected to Wernicke's area by the arcuate fasciculus (a bundle of nerve fibres), Broca's area deals with speech planning and production, controlling both inflectional and functional morphemes. The mechanisms behind semantic interpretation are lateralised throughout the brain from ages 2-3, with Christine Chiarello commenting that, 'Although the right hemisphere seems to retain information about the perceptual characteristics of words, we do know that words presented to either hemisphere are comprehended (although not necessarily in the same way). Hence word meaning access is bilateral.'[5]

Regarding language acquisition (LA) itself, there are many important findings from the psychological and neurobiological literature which can help flesh out an internalist perspective of semantics (which claims that the meaning of word is not determined by the external world, but by internal, syntactic factors, as well as human concerns and interests – more on this shortly). Within six months, for instance, newborn infants can detect language-relevant data from William James' 'blooming buzzing confusion' – no slight task. Many studies have shown that, 'from birth, infants have some capacity to categorize utterances as belonging to different languages according to their rhythmic organization. By 4 months, they can go beyond gross rhythmic distinctions and use other kinds of information to distinguish utterances in their native language from those in another language with a similar rhythmic organization.'[6] The cries of young children also make syntactic distinctions between proper and common nouns.[7] Another factor apparently determining the obscure notion of 'objecthood' (a matter, again, to be returned to shortly) is phonology, with the initial guesses of 2nd-graders about whether a novel word refers to either an action or an object being affected by the specific sound properties of the word.[8]

At six-twelve months old, semantic categories and concepts begin to 'grow' in the mind (using Chomsky's organic metaphor), and at twelve-eighteen months familiar words receive greater LH activity between temporal and frontal lobes. The nine month mark sees a sensitivity in infants to prosodic cues like stress, as they listen more to disyllabic words with STRONGweak stress (FAL.ter) than weakSTRONG (po.LICE), helping them discover predominant stress patterns in 'English.'[9] Other indications suggest that this phonological sensitivity 'may also help learners in certain ways to discover the underlying syntactic organization of utterances.'[10] Infants are also sensitive to phonosyntactic constraints in their I-language, recognising that [br] is generally found at the beginning of a word, and [rt] at the end, aiding the learning of word boundaries.[11]

Of perhaps more traditional 'philosophical' importance is the Homogeneity Hypothesis, advanced by Gary Libben, which maintains that the minds of bilinguals contain a single lexical and conceptual store.[12] A bilingual, in this view, does not have two 'languages,' but rather has interacting lexical items within an integrated lexicon. Lexical organisation itself is based on certain word associations, like semantic field (needle-sew; needle-nail), pairs (black-white), word class (noun-noun), superordination (colour: blue, red) and synonymy (hungry-starved). It is generally agreed among psychologists that L1 acquisition involves greater implicit learning mechanisms (such as bottom-up and automatic processes and a lack of awareness of the result in the learner) than explicit ones. Baars sees learning as a 'magical process,' in the sense that it requires us only to 'point' our attention at some material we want to learn, and learning occurs without any other conscious input, 'carried out by some skilled squad of unconscious helpers.'[13]

In the study of L1 influences on L2 acquisition, Eric Kellerman introduced the notion of 'transferability' (also known as 'psychotypology'). This refers to the claim that L1 transfer is partly a function of learners' (conscious of subconscious) intuitions about how transferable certain phenomena are. The verb 'break,' for instance, can be both transitive and intransitive in Dutch and English. In Kellerman's experiments, Dutch students with partial knowledge of English accepted both uses of the verb 'break' almost 100 per cent of the time. Dutch students with intermediate knowledge of English correctly accepted all the transitive use of 'break' (e.g. 'He broke his leg'), but accepted only 60 per cent of intransitive ones. Dutch students with advanced knowledge of English accepted all transitive items but only 80 per cent of intransitive items. Why would beginners be correct at a higher rate than advanced learners? Kellerman suggests that transitive verbs are semantically more transparent and syntactically more prototypical than intransitive verbs (things do not usually break on their own). More advanced learners develop 'a sensitivity to a pragmatic distinction (implicitly known) between the causative and the noncausative meanings of a single verb.' They go 'beyond success,' with learning being based not just on L1-L2 comparisons but also on human concerns and interests (an important point mainstream 'externalist' philosophy of language typically ignores).[14] Katherine Nelson also discovered in the early 1970s that children who have referential preferences (such as naming things or activities) typically learn more single words, particularly nouns, during LA. Conversely, children who have more expressive and interactional tendencies are more likely to learn whole unsegmented expressions (or 'formulaic language'). The reasons for this may be psycholinguistic, or may reflect what the child 'supposes the language to be useful for.'[15]

What are sometimes called usage-based theories of language acquisition claim that grammatical structures are not genetically determined and result from implicit learning processes triggered by language-relevant data. These theories claim that the starting point of language development is the automatic acquisition of linguistic items. The major usage-based theories (or item-/exemplar-based theories) share the belief that grammatical structures do not come directly from the human genome but are the result of an emergent process both at the macro level of language change throughout the centuries and the micro level of LA. Some form of statistical analysis is doubtless part of LA, but, equally doubtless, is the involvement of what Chomsky calls 'third-factor' considerations, or physical constraints on the development of cognitive capacities (as stressed by the minimalist program). There are many telling similarities between the folk science of Ancient Greece and Rome (where linguistic capacity was ascribed to the tongue), the behaviourism of the 1950s and 1960s and mainstream contemporary cognitive science. Throughout the twentieth century language has widely been seen as a set of behaviours, in the 1960s it was stimulated by conditioned responses, but 'now it's by whatever cognitive processes are around,' for instance 'complicated Bayesian statistics,' Chomsky warns.[16] Even the careful and perceptive mathematician Frank Ramsey fell prey to his behaviourist intuitions by asking his readers to consider the 'mental factors of belief' in the case of 'a chicken [believing] a certain sort of caterpillar to be poisonous. ... The mental factors in such a belief would be parts of the chicken's behaviour, which are somehow related to the objective factors, viz. the kind of caterpillar and poisonousness.'[17]

The major 'non-symbolic psychological theories,' such as connectionism, dynamic systems theory and usage-based theories 'stand in some sort of opposition to the classic symbolic view of cognition and nativism,' and instead 'argue that cognition is an emergent phenomenon, grounded in lower, simpler, non-symbolic processes such as ... the general cognitive capacities that enable humans to establish joint attention, to understand the communicate intention of others, to form categories, to detect patterns, to imitate, to notice novelty, and to have the social drive to interact with conspecific caregivers.'[18] Be that as it may, connectionist and usage-based theories fail to capture even the most elementary aspects of LA, such as learning the regular past tense. For instance the network constructed by McClelland and Rumelhard, 'trained to generalize verbs to their past-tensed forms, correctly predicted *malked* and *splang* for the novel inputs *malk* and *spling*. But it transformed the novel and unsimilar input *ploanth* to *bro*, and *frilg* to *freezled*, something humans would never do.'[19] Contrary to many figures in applied linguistics, philosopher Wolfram Hinzen suggests that 'If human naturally do what connectionist nets have great difficulties in doing, and connectionist nets naturally do what humans do not, then connectionist nets look like a bad model of human nature.'[20]

Considering such broad topics and human nature, in one of his many accounts of Plato's problem Chomsky stresses that 'in rational inquiry, in the natural sciences or elsewhere, there is no such subject as "the study of everything." ... Rather, in rational inquiry we idealize to selected domains in such a way (we hope) as to permit us to discover crucial features of the world.'[21] If we were to ask a physicist to describe what's happening outside the window, they would be hard put to find any answer short of a common sense or 'intentional' descriptions, like 'That man is running' or 'Those cars are moving.' For the natural scientist, there are simply too many factors at work (something which is not, incidentally, a criticism of the strict rationality applied in the sciences – experiments themselves would be impossible without shutting out the hubbub of the world), since 'physical systems aggregate into increasingly complex structures, existing at different levels of organization.'[22] Plato believed intelligibility was to be found only in the world of geometry and mathematics, with the world of sensation an unreal one. An effective study of astronomy, in his view, requires that 'we shall proceed, as we do in geometry, by means of problems, and leave the starry heavens alone.'[23]

In another discussion on the philosophy of science, Chomsky develops this further:

> Science talks about very simple things, and asks hard questions about them. As soon as things become too complex, science can't deal with them. The reason why physics can achieve such depth is that it restricts itself to extremely simple things, abstracted from the complexity of the world. As soon as an atom gets too complicated, maybe helium, they hand it over to chemists. When problems become too complicated for chemists, they hand it over to biologists. Biologists often hand it over to the sociologists, and they hand it over to the historians, and so on. But it's a complicated matter: Science studies what's at the edge of understanding, and what's at the edge of understanding is usually fairly simple. And it rarely reaches human affairs. Human affairs are way too complicated. In fact even understanding insects is an extremely complicated problem in the sciences. So the actual sciences tell us virtually nothing about human affairs.[24]

When it comes to higher mental faculties like language, Chomsky notes that 'it is not unlikely that these are among the most complex structures in the universe.'[25] He also derides Michael Dummett's view that, 'For the sciences, the theories (accuracy aside) tell us everything relevant about the form in which the body of knowledge is delivered; however for the theory of meaning, (and, presumably, language and thought generally, and perhaps vision, reifications, etc.), some additional kind of explanation is required, a "philosophical explanation," that goes beyond science.'[26] But how do 'philosophical accounts,' to use Dennett's phrase, differ from naturalistic ones? Quine tells us that 'the world is as natural science says it is, insofar as natural science is right.'[27] Philosophy, Weinberg stressed, has nothing to say about complex natural laws, like gravity and the second law of thermodynamics. This is true to the same extent that science currently has very little to say about complex phenomena like human affairs. But, turning to a similar approach, Russell believed that philosophical knowledge

> does not differ essentially from scientific knowledge; there is no special source of wisdom which is open to philosophy but not to science, and the results obtained by philosophy are not radically different from those obtained by science. The essential

characteristic of philosophy, which makes it a study distinct from science, is *criticism*.[28]

One of the greatest contemporary philosophers, Galen Strawson, opens one of his essays on metaphysics with the following Russellian statement:

> Philosophy is one of the great sciences of reality. It has the same goal as natural science. Both seek to give true accounts, or the best accounts possible, of how things are in reality. They standardly employ very different methods. Philosophy, unlike natural science, usually works at finding good ways of characterizing how things are without engaging in much empirical or a posteriori investigation of the world. It has a vast field of exercise. Many striking and unobvious facts about the nature of reality can be established a priori, facts about the structure of self-consciousness, for example, or the possibility of free will, or the nature of intentional action, or the viability of the view that there is a fundamental metaphysical distinction between objects and their properties.[29]

Schopenhauer also seems to have respected this idea: 'Philosophy is world-wisdom; its problem is the world.'[30] For Hume, metaphysics is one of the four principal 'branches of science. Mathematics and natural science ... are not half so valuable.'[31] Philosophy is 'the most fascinating topic of all,' for Chomsky.[32] The role of this unique discipline according to Russell was to explore those questions we don't know how to answer; science, by contrast, does the opposite. But though philosophy certainly concerns itself with analysing key terms, this view seems somewhat misleading. Many philosophical questions (especially of the 'existential' type) turn out on inspection to be either meaningless or unanswerable. Take, for instance, 'Why do things happen?' or 'How is nature possible?' Such sentences may have the form of a question, but no substance lies beyond; meaning, it may have the structure of an interrogative expression, but it has no imaginable answers, and so therefore it's not a question.[33]

It's worth noting that, mostly in popular science books, scientists today often begin with an obligatory discussion about what the philosophers have got wrong for so many years, before assuring their audience not to worry – they'll set the record straight. All of this occurs without the author engaging in any of the philosophical issues they easily dismiss (a recent example of this is Stephen Hawking's *The Grand Design*). This is not a criticism of the science discussed: Though sweeping in narrative clarity and impassioned insight, leaving the reader with an ambiguous closing chapter on how to combine the forces of spirituality and science doesn't solve the important problems that have stayed with us for centuries.

A fallacy in the natural sciences since the time of Descartes (strengthened by the logical positivists) has been to assume that evidence about a certain person's nature could only be taken from that particular person's behaviour, and that it is therefore misleading or 'dualistic' to attribute to them an internal mechanism responsible for that behaviour. Such a mechanism would (and in fact does) comprise of certain rules or sets of neural patterning which allow certain behaviour to become realized: 'The a priori framework of human thought, within which language is acquired, provides necessary connections among concepts, reflected in connections of meaning among words and, more broadly, among expressions involving these words.'[34] This rationalist approach sides with Russell's famous comment that a priori knowledge 'cannot be logically deduced from facts of experience. Either, therefore, we know something independent of experience, or science is moonshine.'[35] The British philosopher and mathematician had written earlier that 'Knowledge is called *empirical* when it rests wholly or partly upon experience. Thus all knowledge which asserts existence is empirical, and only *a priori* knowledge concerning existence is hypothetical, giving connexions among things that exist or may exist, but not giving actual existence.'[36] Much has taken place since the time of Kant, when the 'schematism of our understanding, in its application to appearance and their mere form, is an art concealed in the depths of the human soul, whose real modes of activity nature is hardly likely ever to allow us to discover, and to have open to our gaze.'[37] Contemporary ridicule of 'mysterianism' (the claim that human's have cognitive limits, since the mind is a natural object and not a golden ladder to some Platonic realm of 'ideas') is also something which has changed, to the likely disbelief of most Enlightenment figures.

An interesting place to shed some light on the dark depths of Plato's problem is what many have called the hardest problem in science: the evolution of language. If language is, as Leibniz said, 'the best mirror of the human mind,' then any understanding of the mind will need to reserve some space for its mysteries. For Wilhelm von Humboldt, 'the whole mode of *perceiving* things *subjectively* necessarily passes over into cultivation and the use of language.'[38] Though the term 'perception' is often used in philosophy to refer to direct 'veridical' perception (there being a supposed 'representational' causal link between my seeing a cow and a mind-independent object), I will use it simply to refer to sensory experience; the data of sense (what Epictetus called 'impressions') interpreted by sensory servomechanisms.

But it is perhaps a sign of our irrational attitude towards ourselves that much excitement over the evolution of language (of which we know virtually nothing, lack of evidence being the major obstacle) has diverted attention away from problems of the evolution of much simpler organisms (like bees, as Chomsky often points out). Contrary to illusion, even Darwin's *On the Origin of Species* refrained from talking about 'origins' (because the questions are too hard the evidence too sparse) or 'species' (since the term is not atomic and too general for naturalistic inquiry). The hundreds of hardbacks on the evolution of language being published by distinguished university presses may benefit from paying attention to sentiments of Montaigne:

> I have always felt grateful to that girl from Miletus who, seeing the local philosopher ... with his eyes staring upwards, constantly occupied in contemplating the vault of heaven, made him trip over, to warn him that there was time enough to occupy his thoughts with things above the clouds when he had accounted for everything lying before his feet.[39]

In more recent decades, linguists have shown that there are cognitive principles general to humans which determine the form and interpretation of sentences. These 'grow' in the mind as a result of the genetic basis of language (whatever it turns out to be), or 'universal grammar' (UG) in Chomsky's form, originating from his original syntactic theories in the 1950s. UG is the genetically determined 'language faculty' which contains certain principles which structure the grammar of a language learner. These include, roughly, where the verb is put in the sentence, whether the grammar is head-initial (like in English: 'John has put the cup on the table,' but not like in German: 'John-has the-book on-the-table put'), word order (Subject-Verb-Object, Object-Subject-Verb, etc.), and many others. 'Similarly, no child has to learn that there are three and four word sentences, but no three-and-a-half word sentences, and that they go on forever.'[40] Chomsky qualifies that 'A rational Martian scientist would probably find the variation rather superficial, concluding that there is one human language with minor variants.'[41] Similar thoughts are occasionally found in Mill's work, such as during a rectorial address at St Andrews in 1867:

> Consider for a moment what Grammar is. It is the most elementary part of Logic. It is the beginning of the analysis of the thinking process. The principles and rules of grammar are the means by which the forms of language are made to correspond with the universal forms of thought. ... The structure of every sentence is a lesson in logic.[42]

Part of Chomsky's reason for proposing a language faculty was a simple input-output deficiency: 'Compared with the number of sentences that a child can produce or interpret with ease, the number of sentences in a lifetime is ridiculously small.'[43] Understanding a domain-specific mechanism like those which compose the language faculty requires us 'to elucidate not only the formal description of the input domain of a system, but also a complete formal description of its procedures, or what it does with the information that is being processed.'[44] We seem to acquire about a word and hour from the ages two to eight, and have very little choice in the matter – an astonishing feat involving complex statistical analysis and other domain-general mechanisms (since 'without any grammatical instruction, from innumerable sentences heard and understood he will abstract some notion of their structure which is definite enough to guide him in framing sentences of his own, though it is difficult or impossible to state what that notion is except by means of technical terms like subject, verb, etc.'[45]), along with cognitive faculties thought to be part of the language faculty. As the faculty develops, it becomes 'capable of pairing semantic and phonetic interpretations over an indefinite range of sentences to which [we have] never been exposed. Thus [our] knowledge extends far beyond [our] experience.'[46] Linguistic competence appears to be essentially uniform across ethnicities, with Richard Lewontin calculating that 'about 85 per cent of all identified human genetic variation is between any two individuals from the same ethnic group,' with only 7 per cent difference lying between global races from Africa to Australasia.[47] If the language faculty has a genetic component (a virtual necessity for its rapid, universal 'growth'), we would correspondingly expect all manifestations of it (English, Japanese, Bantu, etc.) to be essentially uniform, differing only in trivial and superficial respects (such as characterization and mode of externalization, involving variables such as accent). Today 'explanatory adequacy' continues to be the task of exploring the surface variance of languages to show that 'the apparent richness and diversity of linguistic phenomena is illusory and epiphenomenal, the result of the interaction of fixed principles under slightly varying conditions.'[48]

Whilst outlining his minimalist (biolinguistic) program, Chomsky summarised that 'UG provides a unique computational system, with derivations driven by morphological properties to which syntactic variation of languages is restricted.'[49] He and many biologists argue that the sounds a child hears are far too sparse to account for the rich understanding of their language, and it follows, in the words of the Stanford Encyclopedia of Philosophy, that learners 'have an innate knowledge of a universal grammar capturing the common deep structure of natural languages.'[50] Recent discoveries also suggest that after the initial emergence of modern homo sapiens 200,000kya, a subgroup of humans separated around 150,000kya and were not united again until 50,000kya; meaning that the fundamental computational procedures of language may have been in place as early as 150,000kya. This initial state of 'learning' structures the capacities of this specific 'mental organ' as it 'grows,' like any other system of the body, into the final state or 'I-language.' The nineteenth-century philologists stressed the importance of sounds, in the 1920s Otto Jespersen shifted attention to the written word, and in the 1950s Chomsky moved the study of language into the mind. Speaking on the domain-specific nature of UG, Chomsky recently outlined how 'analytic mechanisms of the language faculty seem to be triggered in much the same ways whether the input is auditory, visual, even tactual, and seem to be localized in the same brain areas, somewhat surprisingly.'[51] There is also good evidence that only a single exposure to lexical items like 'tree' is enough for the meaning to be associated with them. An often ignored implication for the philosophy of language is that the sound-symbol relation is acquired after this process, but the meaning isn't, which is purely internal. Children instead acquire the sensory-motor function for externalizing language, the place where most of the complexity lies:

> Almost everything that's been studied for thousands of years is externalization. When you teach a language, you mostly teach the externalization. Whatever is going on internally, it's not something that we're conscious of. And it's probably very simple. It almost has to be, given the evolutionary conditions.[52]

Hespos and Spelke detail experiments which show that, 'just as in speech perception, children learn not meaning, but which *distinctions* of meaning the language of their environment makes. Specifically, 5-month-old children raised in English-speaking environments are sensitive to a distinction marked in adult Korean but not adult English, namely the distinction between the "tight" or "loose" fit of one object with respect to another.'[53] They conclude that 'the early development of semantic categories parallels the development of phonological categories' and suggest that 'natural language semantics, like natural language phonology, evolved to capitalize on pre-existing representational capacities,' helping us also to parse the world into objects in early infancy.[54] What these pre-existing conceptual capacities are, however, remains in obscurity, with the problem being made no easier by Bloom's conclusion to *How Children Learn the Meanings of Words*: 'Nobody knows how children learn the meaning of words.'[55]

It's almost superfluous, then, to point out that 'No theory in the foreseeable future can hope to account for all of cognition.'[56]

More generally, Gennaro Chierchia explains how our minds are

> constituted by a series of relatively autonomous systems, each using a specific vocabulary and mode of operating. One such system is language, which should be viewed as a sort of "mental organ"; grammar constitutes our inborn knowledge of language. Particular languages arise through the creation of particular lexica and the use of the combinatorial apparatus constitutive of grammar. Such an apparatus allows for a limited number of options, which can be set differently across languages (these options are called the "parameters"). ... The view of intelligence as arising from the interaction of differentiated autonomous components has come to be known as the "modular view."[57]

In an effort to counter this 'modular' view of the brain, in which 'complex functional systems such as attention and memory' are referred to as 'faculties,' Norman Segalowitz writes that these aspects have now 'been fractioned into components that are subserved by different areas of the brain.' Reviewing the work of M. I. Posner and S. E. Peterson, Segalowitz notes that 'anatomically distinct subsystems' have been identified which 'carry out different functions of attention such as orienting to sensory events, detecting signals for conscious processing, maintaining a vigilant or alert state, exerting top-down executive control, and disengaging focus of attention. It is now clear that while regions of the brain carry out highly specialized activities, there is at the same time no single 'module' for attention or consciousness; what we call attention is a complex activity of the whole brain carried out by a highly integrated (in a healthy brain) network of subsystems. The same is true for language.'[58]

But these arguments fail to negate the simple conceptual clarity and even the empirical evidence for modularisation. Segalowitz seems to mistake mechanisms for modules: The language faculty may be disparate in neurological organisation, but this does not deny the modular operation of such mechanisms. A computer may also have its parts strewn across a room, but so long as they are connected their ability for parallel processing is not hindered. For Jackendoff, the brain

> comprises a large number of specialized systems that interact in parallel to build up our understanding of the world and to control our goals and actions in the world. Even what seems to be a unified subsystem such as vision has been found to be subdivided into many smaller interacting systems for detecting

motion, detecting depth, coordinating reaching movements, recognizing faces, and so forth.[59]

Similar things apply to language. Jackendoff again points out how

> Much of the neuroscience of language has been concerned with how words stored in long-term memory are activated ("light up") in the course of sentence perception and production. But activation of words alone is not sufficient to account for the understanding of sentences. If understanding [the sentence "The little star's beside a big star"] consisted only of activating the words, the sentence ["The big star's beside a little star"] ... would "light up" the same words and hence be understood the same.[60]

As Jackendoff has recently stressed, we are only consciously aware through introspection of the phonological aspect of linguistic 'thought,' with the syntactic computations operating unconsciously. Considering the modularity of the two systems, it's not surprising that 'Introspection of one's visual experience does not reveal any intrinsic property of one's experience: what it reveals is what the experience is *about* or what the experience is an experience *of*.'[61] Visual introspection is also metarepresentational and not perceptual.[62] Supporting Jackendoff's view of the autonomous and modular nature of language, Fromkin condenses decades of research as follows:

> The more we look, whether at studies of neonates or development or lesions or blood flow studies of cognitive processes or ERP and fMRI studies, the more we find that knowledge and processing of language is separate from the ability to acquire and process other kinds of knowledge, that the asymmetry between general knowledge and linguistic knowledge shows language to be independent of general intellectual ability, and that language itself, as well as other cognitive systems are distinct both anatomically and functionally.[63]

Many other modules have been argued for, including numerical competence, folk biology and physics, face recognition and morality. Contrary to much popular illusion, many of the Enlightenment empiricists also believed in innate mental structure, though some of them were more reluctant to permit it than others. Don Garrett makes the following observation of Hume:

> Although he identifies the idea of the mind with the idea of a set of bundled perceptions, he is not forbidden by his empiricist

> principles from postulating the existence of unperceived deterministic mechanisms that would underlie the propensities of perceptions to appear in particular ways. He is forbidden by his principles only from trying to specify the nature of those mechanisms beyond what experience can warrant.[64]

Language thus provides 'the framework for the interpretation of experience and the construction of specific forms of knowledge on the basis of experience.'[65] A similar perspective of emotions was given by the celebrated cognitive neuroscientist Antonio Damasio, who writes that 'Emotions provide a natural means for the brain and mind to evaluate the environment within and around the organism, and respond accordingly and adaptively.'[66] Other, adaptationist-inspired views hold that 'the existence of a close relationship between bodily gesture and verbal syntax implies that it is not just concrete nouns, the "thing-words", but even the most apparently formal and logical elements of language, that originate in body and emotion. The deep structure of syntax is founded on the fixed sequences of limb movements in running creatures.'[67] It's also possible that syntax (or the structuralization of lexical items) 'might have derived evolutionarily from the structure of the human syllable, itself deriving from the reconfiguration of the vocal tract due to consistent bipedalism.'[68] This claim is certainly highly contentious, but in terms of its neurophysiology, syntax seems to be connected to parts of the brain involved in hierarchical motor control, like the basal ganglia. It's also been widely speculated that

> the advantages of spoken language, such as more efficient hunting, would have benefitted all individuals in the group, even if there was a range of development of linguistic skills, since the products of hunting would have been shared ... it looks as if language may have started out, not as the product of ruthless competition, but as another area in which humanity has done better by co-operation and collaboration.[69]

The famous and increasingly popular 'embodied' scholars Lakoff and Johnson argue slightly differently that

> the very structure of reason itself comes from the details of our embodiment. The same neural and cognitive mechanisms that allow us to perceive and move around also create our conceptual systems and modes of reason.[70]

This claim, incidentally, is not as 'bold' as many commentators have suggested. Claiming that cognition and language are 'embodied' is a conclusion and standard any biologist would agree to when working on, say, the immune or vascular systems. Claiming that 'philosophy is in the flesh' is in many ways like saying 'photosynthesis is in the chloroplasts.' Humans of course have a 'nature,' like any other organism. The biological implications of innate mental faculties are shocking only to certain philosophers and social 'scientists.' But why many philosophers, linguists and neuroscientists abandon rationality when analysing faculties of the body 'above the neck' and talk in mystic, dualistic, Homeric terms is not the subject of this study (though even the finest psychotherapists would be met with confusion when examining those who stick to such religiously dualistic frameworks). Putting the matter similarly (though agreeing with many of Lakoff's conclusions), Ruth Anshen seems to have something similar in mind:

> Human thought has been formed through centuries of man's consciousness, by perceptions and meanings that relate us to nature. The smallest living entity, be it a molecule or a particle, is at the same time present in the structure of the earth and all its inhabitants, whether human or manifesting themselves in the multiplicity of other forms of life.[71]

Writing in *Science* in 2002, Marc Hauser, William Tecumseh Fitch and Chomsky (in)famously discussed whether 'human language evolved by gradual extension of pre-existing communication systems, or whether important aspects of language have been exapted away from their previous adaptive functions (e.g., spatial or numerical reasoning, Machiavellian social scheming, tool-making).'[72] Most researchers have positioned themselves somewhere between these 'extreme' and 'independent questions.' Spelling out their interdisciplinary biolinguistic framework, they go on to say that

> there has been substantial evolutionary remodeling since we diverged from a common ancestor some 6 million years ago. The empirical challenge is to determine what was inherited unchanged from this common ancestor, what has been subjected to minor modifications, and what (if anything) is qualitatively new. The additional evolutionary challenge is to determine what selectional pressures led to adaptive changes over time and to understand the various constraints that channeled this evolutionary process. ...
>
> The word "language" has highly divergent meanings in different contexts and disciplines. In informal usage, a language is understood as a culturally specific communication system (English, Navajo, etc.). In the varieties of modern linguistics that concern us here, the term 'language' is used quite differently to refer to an internal component of the mind/brain (sometimes called "internal language" or "I-language"). We

assume that this is the primary object of interest for the study of the evolution and function of the language faculty. However, this biologically and individually grounded usage still leaves much open to interpretation (and misunderstanding). For example, a neuroscientist might ask: What components of the human nervous system are recruited in the use of language in its broadest sense? Because any aspect of cognition appears to be, at least in principle, accessible to language, the broadest answer to this question is, probably, "most of it." Even aspects of emotion or cognition not readily verbalized may be influenced by linguistically based thought processes.[73]

Rudimentary syntax also supposedly exists in certain non-primate and non-mammalian systems, such as in the Caribbean reef squid investigated by Jennifer Mather, 'whose skin displays a visual language in which some researchers discern distinctive lists of "grammatical" components, such as nouns, verbs and adjectives.'[74] The I-languages (internal, individual, intensional) Chomsky, Hauser and Fitch investigate are mental organs, analogous to the visual system or immune system, which interact with other sub-systems of the human organism, in mostly unknown ways – part of the full spectrum of actions, decision making, planning and so on.[75] Contrary to the ordinary notion of language as a communication system, the vast majority of language use is mind-internal ('internal dialogue'). We learn early words like 'chair' and 'river' in entirely different ways than how we later learn (after the language faculty is fully developed at around eight or nine years) the words 'vector' and 'amethyst' – they arise through different 'language games,' in Wittgenstein's phrase (dropping the externalist connotations). It's worth noting, as John A. Lucy does, that on top of the numerous language games in the world, 'individual performance and ability' of linguistic traits 'are much the same in every culture,' strengthening the uniformity of the language faculty in this particular respect – or, more specifically, the likelihood of their being a Universal Grammar (an initial state of the brain) inherent in all humans, from which the individual's I-language 'grows,' in the same sense that their intestines or liver does.[76] Or, in the words of Locke, language acquisition involves 'the steady workmanship of nature.'[77] Regarding this linguistic growth, Hinzen speculates quite plausibly that the term 'learning' may in time

> share the fate of the Ancients' terms "air", "earth", or "fire", which were likewise meant to be natural kind terms, but turned out to be human descriptive artefacts that science could make no use of. The same is possible for our terms "belief" and "thought". In their use in epistemology, rationality, theory,

philosophy of mind, and common sense, these terms *are* paradigmatically domain-general, as we can have beliefs or thoughts just about anything, and a belief is partially defined by the general notion of a truth-conditional content.[78]

What have also been termed 'external' or 'public languages' (E-languages), like English, German or Japanese, are quite incoherent concepts (though at the same time, levels of abstraction like 'language' and 'communication' are not illegitimate by assumption – any more than the concept of 'time' is illegitimate to physics, even if all that 'really' exists is random quantum fluctuations – so long as they yield theoretical insights). The distinction is similar to the numeral-number distinction, with the former being external markings and products of behaviour, the latter being mental computations with no direct, Platonic relation to the world. Similarly, there are no English visual or olfactory systems, only individual systems. Under the cognitive and internalist I-language perspective, Jackendoff has recently called the English language 'just some subset of chemically and electrically mediated firings, scattered among huge networks of neurons.'[79] This is perhaps a premature ontological commitment (neurobiological theories are not set in stone), and it doesn't answer the question of what our concept of 'English' is to begin with and how it arises in an individual, but the basic internalist assumptions are certainly on the right track, and will be adopted throughout the rest of this study.

Frege seems to have something externalist in mind when he spoke of the 'common store of thoughts' when outlining his theory of reference.[80] He couched his investigations in a secure E-language framework. As he put it in his classic 1956 paper in *Mind*: 'What does one call a sentence? A series of sounds.'[81] He also 'argued strenuously against treating language (especially meaning) as internal to speakers; his approach underpins most Anglo-American philosophy of language.'[82] Charles Morris, the coiner of the term 'pragmatics,' baptised linguistics in the following way:

> syntax [is] the study of syntactical relations of signs to one another in abstraction from the relations of signs to objects or to interpreters ... semantics deals with the relation of signs to designate and so to objects which they may or do denote ... "pragmatics" is designated the science of the relation of signs to their interpreters.[83]

These E-language illusions are maintained by, for instance, Jerold Katz, who treats language as an abstract Platonic object (to the enduring approval of Paul Postal). John Searle also finds pleasure in telling his undergraduates at Berkeley that what Frege said about language was 'pretty damn' close to the truth, while also equating 'water' with 'H2O' – a mistaken externalist claim to be explored below.[84] My Oxford English Dictionary supplies few words adequate enough to describe how incoherent and confused these views are. But as Galen Strawson has plausibly claimed, the twentieth century was 'the silliest of all the centuries, philosophically speaking (for all its achievements).'[85] Hinzen reminds externalists in his study of minimalism that language

> cannot be a finite set of objects outside our heads (air pressure wave changes, movements of molecules, etc.), given that *it* is not finite; in any case, if it existed outside of us, we would have to assume that internal mechanisms enter into the process of using and knowing it.

Indeed, 'there isn't as much as a *sentence* out there, to be represented by us, without [the existence of internal mechanisms].'[86] These ideas have rich historical antecedents, with the I/E-language distinction directly mirroring Alonzo Church's terminology for his interpretation of the lambda-calculus.[87] This biolinguistic (I-language) definition of language is as vital to cognitive science as spatiotemporal notions were to Einstein. With striking implications for the philosophy of language, Hinzen follows the Cartesian-rationalist method reinvoked by the biolinguistic enterprise in stating that 'Factual utterances and public language uses are no more than effects [of the generative rule systems of language] ... to be deduced from a highly abstract account of linguistic structure, built up from a small number of explanatory principles and conceptual necessities.'[88] The importance of idealising to certain domains of inquiry also seems to be implied in Zeno Pylyshyn's discussion of the visual system: 'To use the term "vision" to include all the organism's intellectual activity that originates with information at the eye and culminates in beliefs about the world, or even actions is not very useful, since it runs together a lot of different processes.'[89] But only language allows us to distinguish between types (e.g. 'car') and tokens (e.g. 'Mercedes') – vision does not. Although animals can also categorise, only language can 'attend to relations among categories. ... No other sorts of imagery permits onto explicitly attend to *lack* of information about the world, as expressed by questions: *Is there a doctor in the house?*'[90]

In his history of the generative enterprise, Richard Kayne reminds us that 'it is entirely likely that no two speakers of English have exactly the same syntactic judgements' – with the same applying for our biologically determined moral judgements.[91] As Norbert Hornstein and Louise M. Antony have also written, 'It is not that speakers communicate because they have an E-language in common; rather, where I-languages overlap sufficiently, communication is possible.'[92] This is likely the most crucial and misunderstood aspect of language – namely, that a language is purely mind-internal, like our sense of pain, morals, and aesthetic preferences. Much debate in the current philosophical literature seems to result from a confusion of the implications behind this. There is no 'English language' in the world, only individual grammars which learn words at different rates, in different environments, and consequently associate a vast number of concepts with different lexical items (one man's 'John' is another man's 'toilet'). Donald Davidson reminds us that 'we all talk so frequently about language, or languages, that we tend to forget that there are no such things in the world; there are only people and their various written and acoustical products. This point, obvious in itself, is nevertheless easy to forget.'[93] Talk of the English language, adds Jackendoff, is often a 'harmless reification of the commonality in the linguistic [knowledge] of a perceived community of speakers.'[94] Reifying 'languages' is similar, then, to taxonomising species – to be used for the purposes of empirical inquiry, with no ontological presuppositions or commitments attached. This 'Galilean' naturalism in the mind sciences is a strictly methodological stance with no ontological questions of "materialism" or "physicalism" arising. The language faculty is also a highly integrated system in the brain: we can no more 'take out' the language faculty from the body any more than we can 'take out' the immune system. Berwick, Pietroski, Yankama and Chomsky recently commented how 'Focusing on eye organization and ontogenetic maturation, as opposed to environmental variation – taking the target of inquiry to be species-invariant internal aspects of organisms – has led to clearer understanding in studies of vision. Many cases in biology fit this pattern.'[95]

Paul Horwich has described the internalist (or mentalist) theory of language in this concise summary, in which he details how the biolinguistic program assumes

> (a) that each human being indeed has a faculty of language, FL, a component of his mind-brain constituting the primary causal/explanatory basis of his linguistic activity; (b) that the possible states, $L1, L2, \ldots$, of FL are, by definition, possible I-languages; (c) that each such state, L, is a computational procedure that generates infinitely many I-expressions, $E1, E2, \ldots$; (d) that each such expression, E, is a pairing <PHON(E), SEM(E)> of phonetic and semantic objects, which, through their

respective interaction with the perceptual/articulatory system (P/A) and the conceptual/intentional system (C/I), determine an association of a sound with a thought; (e) that these PHON-SEM pairs are constructed from lexical items, LI1, LI2, ...; and (f) that these lexical items are stored in a lexicon which is accessed by the computational procedures that form I-expressions.[96]

What is meant by the 'conceptual/intentional system' is highly debatable (in fact even calling it debatable is debatable, due to a lack understanding in neurobiology about the evolution of cognition). What we do assume, though, is that the language faculty 'is a computational system (narrow syntax) that generates internal representations and maps them into the sensory-motor interface by the phonological system, and into the conceptual-intentional interface by the (formal) semantic system.'[97] As a result of an individual's UG, 'he is capable of pairing semantic and phonetic interpretations over an indefinite range of sentences to which he has never been exposed. Thus his knowledge extends far beyond his experience and is not a "generalisation."'[98] The language faculty takes a 'finite set of elements and yields a potentially infinite array of discrete expressions' – or, as Wilhelm von Humboldt put it, language is a function which 'must therefore make infinite use of finite means, and is able to do so through the productive power that is the identity of language and thought.'[99] For Colin McGinn, 'meaning is essentially freedom from the actual; a sentence can mean something that has no counterpart in the real world.' Like imagination, meaning 'is a matter of *alternatives*.'[100] Having a similar discussion, a comedy sketch by Stephen Fry and Hugh Laurie helps illustrate this creative capacity:

> Imagine a piano keyboard, eh, 88 keys, only 88 and yet, oh and yet, hundreds of new melodies, new tunes, new harmonies are being composed upon hundreds of different keyboards every day in Dorset alone. Our language, tiger, our language: hundreds of thousands of available words, frillions of legitimate new ideas, so that I can say the following sentence and be utterly sure that nobody has ever said it before in the history of human communication: "Hold the newsreader's nose squarely, waiter, or friendly milk will countermand my trousers." Perfectly ordinary words, but never before put in that precise order. A unique child delivered of a unique mother.[101]

Dr Johnson famously standardised the English language, but one cannot begin to standardise the human mind. With the human mouth only being able to produce around ninety phonemes, it's easy to draw the same conclusion as Humboldt and Fry. Limitations to the sensory-motor system and the working memory of the individual also impose constraints on the complexity of sentences. Hauser, Chomsky and Fitch have concluded:

> The computational system must (i) construct an infinite array of internal expressions from the finite resources of the conceptual-intentional system, and (ii) provide the means to externalize and interpret them at the sensory-motor end. We may now ask to what extent the computational system is optimal, meeting natural conditions of efficient computation such as minimal search and no backtracking. To the extent that this can be established, we will be able to go beyond the (extremely difficult, and still distant) accomplishment of finding the principles of the faculty of language, to an understanding of why the faculty follows these particular principles and not others. We would then understand why languages of a certain class are attainable, whereas other imaginable languages are impossible to learn and sustain. Such progress would not only open the door to a greatly simplified and empirically more tractable evolutionary approach to the faculty of language, but might also be more generally applicable to domains beyond language in a wide range of species – perhaps especially in the domain of spatial navigation and foraging, where problems of optimal search are relevant.[102]

This biolinguistic view of language is sympathetic to the way Humboldt construed it. He stressed how language is not a product (*Ergon*) but an activity (*Energeia*), what linguists today would call syntactic derivations and not the 'work' of externalizing them.[103] The importance of the above biolinguistic, computational approach to the mind should not be overlooked, with Dennis Hassabis suggesting that 'from a neuroscience perspective, attempting to distil intelligence into an algorithmic construct may prove to be the best path to understanding some of the enduring mysteries of our minds, such as consciousness and dreams.'[104] Our linguistic and thought processes grow out of the natural environment and the properties of the brain handed to us by Hume's 'original hand of nature' – either that or, like Adam, they come from 'nowhere.'

> But though animals learn many parts of their knowledge from observation, there are also many parts of it which they derive from the original hand of nature; which much exceed the share of capacity they possess on ordinary occasions; and in which they improve, little or nothing, by the longest practice of

> experience. These we denominate Instincts, and are so apt to admire as something very extraordinary, and inexplicable by all the disquisitions of human understanding. But our wonder will, perhaps, cease or diminish, when we consider, that the experimental reasoning itself, which we posses in common with beasts, and on which the whole conduct of life depends, is nothing but a species of instinct or mechanical power, that acts in us unknown to ourselves; and in its chief operations, is not directed by any such relations or comparisons of ideas, as are the proper objects of our intellectual faculties. Though the instinct be different, yet still it is an instinct, which teaches man to avoid the fire; as much as that, which teaches a bird, with such exactness, the art of incubation, and the whole economy and order of its nursery.[105]

So does the English language exist in the same way that red blood cells do? Naturalistic theories of I-languages simply do not account for the incoherent idea of a 'public language' (English, French, German), or E-language (external) we so often talk about, namely some Platonic notion to which speakers stand in cognitive relation to:

> we can think of the initial state of the language faculty as a fixed network connected to a switch box; the network is constituted of the principles of language, while the switches are the options to be determined by experience. When the switches are set one way, we have Swahili; when they are set another way we have Japanese. Each possible human language is identified as a particular setting of the switches – a setting of parameters, in technical terminology.[106]

But it might turn out that there is no language faculty at all, only some common principles of imitation which allow the sequencing of phonetic constructs and map them to certain areas of the conceptual-intentional system – ideas which have gained in popularity since the early 1990s: 'A cross-cutting division of the initial state [of the language faculty] is between those aspects that belong to a cognitive specialization for language learning and those that belong to more general faculties of the [mind], such as sociability, ability to conceptualize the world, rhythmic analysis of temporal signals, and the ability to form hierarchical structures.'[107] This hierarchy and asymmetry is an elementary aspect of syntax: The construction 'John loves Mary' is not 'flat,' so to speak. Rather, 'Mary' has a more local relation to the verb, a relation which can be expressed as [[John] [loves Mary]] as opposed to [[John] [loves] [Mary]]. For Aristotle, 'The instinct for imitation is inherent in man from his earliest days; he differs from other animals in that he is the most imitative of creatures and he learns his earliest lessons from imitation.'[108] Our large vocabulary size, writes Cedric Boeckx, 'is likely to be tied to our ability to imitate – to unique to us, but not specific to language.'[109] Newborn infants only 42 minutes old can also replicate adult gestures on mouth-opening, tongue-protrusion and lip-protrusion, flexing the mechanisms used for language growth.[110]

But it remains clear that there are human-specific elements of the current theories of Universal Grammar which allow infants to detect human words amongst the mass of sounds around them, and that our knowledge is not based on associative learning or stimulus-response mechanisms. The architecture of the interface components (syntactic and phonological) is likely to form part of UG. In a review of the two 'cognitive revolutions' of the seventeenth century and the 1950s, Chomsky puts this proposition in context:

> In sharp contrast to the rationalist view, we have the classical empiricist assumption that what is innate is (1) certain elementary mechanisms of peripheral processing (a receptor system), and (2) certain analytical mechanisms or inductive principles or mechanisms of associations. What is assumed is that a preliminary analysis of experience is provided by the peripheral processing mechanisms and that one's concepts and knowledge, beyond this, are acquired by application of the innate inductive principles to this initially analyzed experience. Thus only the procedures and mechanisms for acquisition of knowledge constitute an innate property.[111]

It's also highly significant that all other species on the planet manage to communicate just fine without recourse to something as divergent and complex as human language. Even bacteria have some form of communication:

> Many social animals have some system of communication by signs and signals, but language is a species-specific, exclusive property of man. Even "Mongolian" idiots, incapable of looking after themselves in most primitive ways, are capable of acquiring the rudiments of speech – but not dolphins and chimpanzees, highly intelligent as they are in other respects.[112]

Gestures, too, are primary modes of communication throughout the animal kingdom. Humans make good use of hand gestures, for instance, something which might bring one to conclude, like John Napier, that 'if language was given to men to conceal their thoughts, then gesture's purpose was to disclose them.'[113] Or, as Sapir put it, with equal accuracy: 'We respond to gestures with an extreme alertness and, one might also say, in accordance with an elaborate and secret code that is written nowhere, known by none, and understood by all.'[114]

But as Iain McGilchrist writes, it's hard to imagine communication working any other way: 'The realm of all that remains, and has to remain, implicit and ambiguous is extensive, and is crucially important. That is why one feels so hopeless relying on the written word to convey meaning in humanly important and emotionally freighted situations.'[115] Like the spine, or any other part of the body, language is not a functionally 'perfect system' – such things do not exist in nature (though its basic computational procedure, recursion, may be 'perfect' and 'elegant' in the physicist's sense). Though evolution may be a 'tinkerer,' doing the best it can with the material available, physical constraints (which structure the 'canalization' of organic development, to use Waddington's term) also playing a vital role, as minimalism (the latest phase of generative grammar) stresses – the theme of the next chapter. In 1987 V. S. Rotenberg and V. V. Arshavsky wrote of the brain's two hemispheres that

> Non-verbal behaviour, language, facial expression, intonations and gestures are instrumental in establishing complex contradictory, predominantly emotional relations between people and between man and the world. How frequently a touch by the shoulder, a handshake or a look tell more than can be expressed in a long monologue. Not because our speech is not accurate enough. Just the contrary. It is precisely its accuracy and definiteness that make speech unsuited for expressing what is too complex, changeful and ambiguous.[116]

The Chicago-based Canadian writer Nathanaël (formerly Natalie Stephens) often expresses similar feelings in her poetry and prose, 'for I am not convinced I write in any language, nor even that this language is capable of expressing both my disorientation and my effusion. ... However much one attempts to calculate all the cells in the body, it remains nonetheless that in the end, there is you across from me, our eyes cast aside. ... How stupid to think oneself moveable, we don't move, we are nailed to the mouths of others.'[117] For Friedrich Lange, by using language to describe objects 'The name is made a thing, but a thing having no similarity with any other thing, and to which, in the nature of human thought, only negative predicates can be attached. But since there is an absolute necessity for some positive assertion, we find ourselves from the outset in the region of myth and symbol.'[118] This kind of open secrecy was described by Dickens in the following way:

> A wonderful fact to reflect upon, that every human creature is constituted to be that profound secret and mystery to every other. A solemn consideration, when I enter a great city by night, that every one of those darkly clustered houses encloses its own secret, that every room in every one of them encloses its own secret; that every beating heart in the hundreds of thousands of breasts there, is, in some of its imaginings, a secret to the heart nearest it![119]

Isaiah Berlin perhaps goes further:

> The notion of depth is something with which philosophers seldom deal. Nevertheless it is a concept perfectly susceptible to treatment and indeed one of the most important categories we use. When we say that a work is profound or deep, quite apart from the fact that this is obviously a metaphor, I suppose from wells, which are profound and deep – when one says that someone is a profound writer or that a picture of a work of music is profound, it is not very clear what we mean, but we certainly do not wish to exchange these descriptions for some other term such as "beautiful" or "important" or "construed according to rules" or even "immortal" ... According to the romantics – and this is one of their principal contributions to understanding in general – what I mean by depth, although they do not discuss it under that name, is inexhaustibility, unembraceability ... in the case of a work that is profound the more I say the more remains to be said. There is no doubt that, although I attempt to describe what their profundity consists in, as soon as I speak it becomes quite clear that, no matter how long I speak, new chasms open. No matter what I say I always have to leave three dots at the end ... I am forced in my discussion, forced in description, to use language which is in principle, not only today but for ever, inadequate for its purpose.[120]

The primary function of language (note: not it's 'purpose,' quite an incoherent notion in a world lacking a divine overseer) is internal dialogue and the organisation of one's thoughts which can then be externalised for interpretation by conspecifics. The origins of language go too far back in human prehistory to give a definite understanding as to how it developed, and there are no comparable biological systems. Language is 'first and foremost a combinatorial system' according to Elizabeth Spelke, which 'may serve as a medium for constructing new concepts once words and expressions are linked to representations from multiple core systems. These combinations, in turn, make available a new range of potential actions.' Spelke goes on to say how an emerging view of human cognition and action suggests that

> humans are endowed with a set of core knowledge systems, each with its own evolutionary history. These systems may give us the building-block concepts – like *food*, *plate*, *person*, *long*, *left*, and *blue* – that we assemble into thoughts. ... As children learn a natural language, this system may allow them freely to combine their existing concepts and form new ones.
>
> Although the concepts and thoughts provided by core knowledge systems may be limited and fixed, the concepts humans construct with our combinatorial system are more numerous and flexible, and they are not directly constrained by our evolutionary history. Concepts like *food in a blue dish* or *left of a blue wall* may have been useless to ancestral humans, and we do not appear to have evolved any domain-specific cognitive system that expresses them. Our combinatorial capacity nevertheless makes these concepts, and indefinitely many other concepts, available to us.[121]

'Language' is consequently more of a metaphor to describe the 'end state' or I-language of an individual human. As Descartes demonstrated, the abstract mental frameworks which allow us to recognise and imagine triangles and squares are, like language, part of the innate architecture of the mind which allow the identification and manipulation of objects. The claims of Donald Davidson, though not entirely false, lead us into a slightly different and equally as confused discussion:

> As Ludwig Wittgenstein, not to mention Dewey, G. H. Mead, Quine, and many others have insisted, language is intrinsically social. This does not entail that truth and meaning can be defined in terms of observable behaviour, or that it is 'nothing but' observable behaviour; but it does imply that meaning is entirely determined by observable behaviour, even readily observable behaviour. That meanings are decipherable is not a matter of luck; public availability is a constitutive aspect of language.[122]

A decade earlier, Chomsky had written more perceptively that it

> is frequently alleged that the function of language is communication, that its "essential purpose" is to enable people to communicate with one another. It is further alleged that only by attending to the essential purpose can we make sense of the nature of language. It is not easy to evaluate this contention.

> What does it mean to say that language has an "essential purpose"? Suppose that in the quiet of my study I think about a problem, using language, and even write down what I think. Suppose that someone speaks honestly, merely out of a sense of integrity, fully aware that his audience will refuse to comprehend or even consider what he is saying. Consider informal conversation conducted for the sole purpose of maintaining casual friendly relations, with no particular concern as to its content. Are these examples of "communication"? If so, what do we mean by "communication" in the absence of an audience, or with an audience assumed to be completely unresponsive, or with no intention to convey information or modify belief or attitude?[123]

For Russell, too, 'The chief importance of knowledge by description is that it enables us to pass beyond the limits of our private experience.'[124] To demonstrate the inefficiency of language's attempt at efficient 'communication,' we can draw on one of Chomsky's examples of a violation of the Empty Category Principle. Taking the sentence 'He wondered whether the mechanics fixed the cars,' we can ask two questions of it: First, the grammatical sentence 'How many cars did he wonder whether the mechanics fixed?', and second, the ungrammatical sentence 'How many mechanics did he wonder whether fixed the cars?' (though this is 'a fine thought,' as Chomsky points out, we just can't express it that way). There is something about the design of language which acts a barrier to communication, preferring to act as a useful 'thought system' instead. Taking this into account, we can nod with Hinzen when he suggests that 'As a communication device ... [language] is probably flawed in crucial respects, containing lots of structure that seems sub-optimal, redundant, or worse.'[125]

This points to a further question sometimes raised by James Higginbotham, namely the fact that language is internally hierarchical and yet is externalized one-dimensionally through speech (following the escalating tradition in linguistics of naming problems after people, we can call this 'Higginbotham's problem'). Johan Haman seems to have hinted at the nature of this phenomenon, writing that 'To speak is to translate – from a language of angels into a language of men.'[126] As Higginbotham also points out, strings of words can be simultaneously comprehensible and defective.[127] To use a recent example of Berwick, Pietroski, Yankama and Chomsky's, a speaker who would never use '1' can still know it means '2' and not '2a':

1. * The child seems sleeping.
2. The child seems to be sleeping.
2a. The child seems sleepy.

They note how 'Our capacity to understand [1] is no more mere capacity to "repair" the string by converting it into an expression of English, since [2] is well-formed and similar to [2]. Descriptions of linguistic competence must accommodate such facts.'[128] The narrow faculty of language (FLN, possibly just Merge), then, seems to fit Haman's 'language of angels' more than its modality-independent mode of externalization: 'linear ordering of words may reflect *externalization* of linguistic structures by the sensory-motor system, as opposed to the generative procedures under consideration here: While $\{Z, \{X, Y\}\}$ differs from $\{Y, \{X, Z\}\}$ in ways suggestive of how *Bob saw Al* differs from *Al saw Bob*, mere words order seems to be irrelevant for semantic composition.'[129]

In the launch title of the *Oxford Studies in Biolinguistics* series, *The Biolinguistic Enterprise*, Robert Berwick and Chomsky examine this furore over communication:

> The inference of a biological trait's 'purpose' or 'function' from its surface form is always rife with difficulties. Lewontin's remarks in *The Triple Helix* illustrate how difficult it can be to assign a unique function to an organ or trait even in the case of what at first seems like a far simpler situation: bones do not have a single, unambiguous function. While it is true that bones support the body, allowing us to stand up and walk, they are also a storehouse for calcium and bone marrow for producing new red blood cells, so they are in a sense part of the circulatory system.[130]

Communication is a more-or-less affair: if our sentences match closely enough to our internal grammars then some kind of communication occurs between us, but it's fairly ancillary to language altogether. This also seems to be the case with concepts, which are 'tuned' to a 'joint intention' between speakers, to use Searle's terminology.[131] This intuitive sense of convergence with other's views of the world also gives us the impression we hold an epistemologically objective view. Heightened awareness of our own subjectivity arise when these world views conflict. Humboldt seems to have touched upon similar observations:

> [People] do not understand one another by actually exchanging signs for things, nor by mutually occasioning one another to produce exactly and completely the same concept; they do it by touching in one another the same link in the chain of their sensory ideas and internal conceptualizations, by striking the

same note on their mental instrument, whereupon matching but not identical concepts are engendered in each.[132]

This is something Somerset Maugham might have agreed with, since he thought that 'if nobody spoke unless he had something to say ... the human race would very soon lose the use of speech.'[133] J. Hughlings Jackson also observed how 'We speak, not only to tell others what we think, but to tell ourselves what we think.'[134] 'Communication,' write Tony Stone and Martin Davies, 'does not require that there should be shared public meanings any more than it requires that there should be shared public pronunciations.'[135] Language is not primarily for communication, Humboldt thought, because 'Man is not so needy – and inarticulate sounds would suffice for the rendering of assistance.'[136] Chomsky interprets François Jacob as holding similar opinions:

> "The quality of language that makes it unique does not seem to be so much its role in communicating directives for action" or other common features of animal communication, Jacob continued, but rather "its role in symbolizing, in evoking cognitive images," in "molding" our notion of reality and yielding our capacity for thought and planning, through its unique property of allowing "infinite combinations of symbols" and therefore "mental creation of possible worlds," ideas that trace back to the seventeenth-century cognitive revolution.[137]

Contrary to the prim 'defenders of grammar' who hiss at those ignorant of the distinction between 'less' and 'fewer,' Chomsky asks 'Is there any other (aside from mis-using technical terms) concepts of "misuse of language?" I am aware of none.'[138] Even Chaucer often used double-negatives, today considered characteristic of 'non-standard English.' And though it's true language is used to talk about the world as conceptualised by the individual (as Jackendoff stresses), the level of Humboldtean creativity and Cartesian freedom the system has appears to have no analogue in the organic world:

> there do not seem to be constraints on [talking about the world] as specific as those we expect in a system detecting object permanence. There is no selective advantage in changing language to cope with *particular* conditions, an observation leading to doubts with respect to the project of assigning language an adaptive function and explaining it in those terms. In short, not only might there also have been enough time to tinker with the language faculty, there might also have been no evolutionary *point* in changing it. ... The weaker the connection of language is to the outside world, the more stable we expect language to be across changes in environmental conditions.[139]

So far as it's possible to tell, the vast majority of language use is internal to the mind (talking to oneself, planning, etc.), with only a fraction externalised (and a fraction of that externalised for the purpose of 'communication'). Language has, throughout history, mostly been used to record natural events, express one's thoughts, and refer to things (hence Chomsky's Romantic belief that language is for 'beauty' rather than 'communication'). Hinzen makes the crucial point (not even conceived by mainstream evolutionary psychologists) that 'Whether *communication* is a distinct function additional to [the various aspects of language use], or simply an abstraction denoting one joint overall effect of several of them on certain occasions, is unclear.'[140] Along with Hinzen, considering language's various roles biolinguistics operates on the assumption that thought is primary, externalization secondary, and communication tertiary (even speech itself involves features primarily evolved for respiration or ingestion). It is odd, then, when many cases of language impairment are referred to as 'communication disorders.' The case of mental health disorders perhaps also suggests the low importance communication has to language, since

> if you listen carefully even to acutely disturbed schizophrenic persons, what they say does make some sense when looked at as an expression of their own individuality and their own attempt to interpret the world around them: for example, ideas that are apparently paranoid and delusional may indeed spring from the schizophrenic's hypersensitivity to interpersonal events which, while certainly misinterpreted, may nevertheless have a basis in reality.[141]

Berwick and Chomsky also survey recent psycholinguistic studies, commenting that 'externalization appears to be modality-independent, as has been learned from studies of sign language in recent years. The structural properties of sign and spoken language are remarkably similar. Additionally, acquisition follows the same course in both, and neural localization seems to be similar as well. That tends to reinforce the conclusion that language is optimized for the system of thought, with mode of externalization secondary.'[142] It's also been speculated that if the language faculty were re-wired, connecting to particular motor systems, it may even be possible for a subject to externalise language by moving his legs in particular motions. This may not be so surprising, considering the secondary role externalisation plays, and considering that language has even been acquired by people 'with no sensory input beyond what can be gained by placing one's hand on another person's face and throat' – as in the case of the deafblind Helen Keller.[143] Drawing also on paleoanthropological evidence, Berwick and Chomsky quote Ian Tattersall's comments that

> "a vocal tract capable of producing the sounds of articulate speech" existed over half a million years before there is any evidence that our ancestors were using language. "We have to conclude," he writes, "that the appearance of language and its anatomical correlates was not driven by natural selection, however beneficial these innovations may appear in hindsight" – a conclusion which raises no problems for standard evolutionary biology, contrary to illusions in popular literature.[144]

Even the fabled FOXP2 gene (or the 'language gene,' as it's been erroneously described in major scientific publications, in particular *Nature*) seems to be part of the process of externalization – or the 'performance' side of language (neuromuscular activity responsible for production and comprehension), not the 'competence' side (or neuroanatomy, with the competence-performance distinction being parallel to Claude Bernard's anatomy-physiology distinction):

> Recent discoveries in birds and mice over the past few years point to an 'emerging consensus' that this transcription factor gene is not so much part of a blueprint for internal syntax, the narrow faculty of language, and most certainly not some hypothetical 'language gene' (just as there are no single genes for eye color or autism) but rather part of regulatory machinery related to externalization. FOXP2 aids in the development of serial fine motor control, orofacial or otherwise: the ability to literally put one sound or gesture down in place, one point after another in time.[145]

At least 150 other genes are implicated in language, so FOXP2 is far from the most vital component.[146] Other genes which may play a crucial role in the development of the language faculty are

> the genes ASPM and microcephalin, which play an important role in increased brain size of our species, though not, apparently, in controlling brain-size differences within the species. Surprisingly, variants of these genes have been found to co-vary with a linguistic trait: the likelihood of tonal languages (specifically the use of pitch to convey lexical or grammatical distinctions), across the globe. Although this story remains to be worked out, this correlation raises the possibility of a causal relationship between within-species genetic variation and linguistic variation, which, if true, would violate one of the most

basic assumptions in linguistic theory (and human biology more generally): that all normal humans could learn any of the world's languages with precisely equal ease.[147]

In a recent essay on the complexities behind the genetic influence on 'language evolution' (quite a misleading term, incidentally, since I-language itself doesn't evolve, though language *use* certainly does), Diller and Cann present the following picture of current genetic research:

> One intriguing target of *FOXP2* is *CNTNAP2*, a gene involved in cortical development and axonal function. High levels of *CNTNAP2* have been found in language related circuits. Polymorphisms in *CNTNAP2* have been associated with specific language impairment (as tested by nonsense word repetition), and with language delays in children with autism. [It's also been] found that FOXP2 protein binds to and directly downregulates *CNTNAP2*, so that in the developing human cortex, the lamina that contain the most FoxP2 protein have the lowest levels of *CNTNAP2*. But if *FOXP2* and *CNTNAP2* are negatively correlated, how can high levels of both be important for language? This simple finding demonstrates the complexity of the gene networks regulated by *FOXP2*.[148]

Kuhl and Miller's findings also suggest that chinchillas perceive stop consonants in a similar way to humans.[149] The perceptual-articulatory mechanisms may be similar, but the conceptual systems, not surprisingly, appear to differ wildly. Other advances in neural imaging have revealed that up to ninety percent of communication is non-verbal, mostly executed through intonation, stress, body language, eye contact, and a host of other factors. An original adaptationist ('just-so') account of the rise of language, though lacking any evidence, holds that:

> For our primate ancestors, who clearly had no speech, body language played a vital role in social cohesion, especially in prolonged sessions of mutual grooming. One theory is that singing, a sort of instinctive musical language of intonation, came into being precisely because, with the advent of humans, social groups became too large for grooming to be practical as a means of bonding. Music, on this account, is a sort of grooming at a distance; no longer necessitating physical touch, but a body language all the same. And, the theory goes, referential language was a late evolution from this.[150]

The notion of 'reference' (and adaptationism) in language will be covered below, but Coetzee's words add a final note of contrast: 'His own opinion, which he did not air, is that the origins of speech lie in song, and the origins of song in the need to fill out with sound the overlarge and rather empty human soul.'[151]

REFERENCES

[1] Aldous Huxley, *Music at Night and Other Essays* (New York: Doubleday, Doran & Co., 1931), 'Sermons in Cats.'
[2] George Davie, *Democratic Intellect* (Edinburgh University Press, 1961), p. 274 (emphasis his), cited in Noam Chomsky, *Deterring Democracy* (London: Vintage, 1992), p. 251.
[3] Bertrand Russell, cited in Noam Chomsky, *Problems of Knowledge and Freedom: The Russell Lectures* (London: Barrie & Jenkins, 1972), p. 43.
[4] Arthur Eddington, *The Nature of the Physical World* (New York: Macmillan, 1928), p. 276.
[5] Christine Chiarello, 'Parallel Systems for Processing Language: Hemispheric Complementarity in the Normal Brain,' *Mind, Brain, and Language*, p. 236 (229-47).
[6] Peter W. Jusczyk, 'The Role of Speech Perception Capacities in Early Language Acquisition,' *Mind, Brain, and Language: Multidisciplinary Perspectives*, ed. Marie T. Banich and molly Mack (London: Lawrence Erlbaum Associates, 2003), p. 64 (61-83).
[7] John Macnamara, *Names for Things* (MIT Press, 1982).
[8] S. A. Fitnevo, M. H. Chrsitiansen and P. Monghan, 'From sound to syntax: Phonological constraints in children's lexical categorization of new words,' *Journal of Child Language*, 36(5), 2009: 967-997.
[9] Peter Juszcyk, Angela D. Friederici, Jeanine M. I. Wessels, Vigdis Y. Svenkerud, and Marie Jusczyk, 'Infants' sensitivity to the sound patterns of native language words,' *Journal of Memory and Language*, vol. 32, no. 3, 1993: 402-420.
[10] Jusczyk, *Mind, Brain, and Language*, p. 79.
[11] A. D. Friederici and J. M. I. Wessels, 'Phonotactic Knowledge of Word Boundaries and its Use in Infant Speech Perception,' *Perception and Psychophysics*, vol. 54: 287-95.
[12] Gary Libben, 'Representation and Processing in the Second Language Lexicon: the Homogeneity Hypothesis,' in John Archibald (ed.), *Second Language Acquisition and Linguistic Theory* (London: Blackwell, 2000), pp. 228-48.
[13] B. J. Baars, *In the Theater of Consciousness: The Workspace of the Mind* (New York: Oxford University Press, 1997), p. 304.
[14] Eric Kellerman, 'An Eye for an Eye: Crosslinguistic Contraints on the Development of the L2 Lexicon,' in Eric Kellerman and M. Sharwood Smith (eds.), *Crosslinguistic Influence in Second Language Acquisition* (New York: Pergamon Press, 1986).
[15] Katherine Nelson, 'Structure and strategy in learning to talk,' *Monographs of the Society for Research in Child Development*, vol. 38, (1 & 2, Serial no.149).
[16] Noam Chomsky, 'Grammar, Mind and Body – A Personal View,' Lecture at Maryland University, 26th January 2012, part of the 2011-12 Dean's Lecture Series: http://www.youtube.com/watch?v=wMQS3klG3N0.
[17] Frank P. Ramsey, 'Facts and Propositions,' *The Foundations of Mathematics*, ed. R. B. Braithwaite (London: Routledge & Kegan Paul, 1931), p. 144.
[18] Zoltán Dörnyei, *The Psychology of Second Language Acquisition* (Oxford University Press, 2008), p. 88.

[19] Wolfram Hinzen, *Mind Design and Minimal Syntax* (Oxford University Press, 2006), p. 49. See J. L. McClelland, D. E. Rumelhard and the PDP Research Group (eds.), *Parallel Distributed Processing: Explorations in the Microstructure of Cognition*, vol. 2 (MIT Press, 1986).
[20] Ibid., p. 50,
[21] Noam Chomsky, *New Horizons in the Study of Language and Mind* (Cambridge University Press, 2002), p. 49.
[22] Jeffrey Yoshimi, 'Supervenience, Dynamical Systems Theory, and Non-Reductive Physicalism,' *The British Journal for the Philosophy of Science*, 63(20), June 2012: 373 (373-98).
[23] Plato, *The Republic of Plato*, trans. F. M. Cornford (Oxford University Press, 1945), pp. 248-9.
[24] 'Science in the Dock: Discussion with Noam Chomsky, Lawrence Krauss & Sean M. Carroll,' *Science & Theology News*, 1 March 2006.
[25] Cited in Randy Allen Harris, *The Linguistics Wars* (Oxford University Press, 2003), p. 65.
[26] Chomsky, *New Horizons in the Study of Language and Mind*, p. 94.
[27] W. V. O. Quine, 'Structure and nature,' *Journal of Philosophy*, 89: 5-9, 1992.
[28] Bertrand Russell, *The Problems of Philosophy* (Oxford University Press, 1959/1912), p. 87 (emphasis his).
[29] Galen Strawson, *Real Materialism and Other Essays* (Oxford University Press, 2008), p. 1.
[30] Arthur Schopenhauer, *The World as Will and Representation*, trans. E. J. F. Payne (New York: Dover, 1969/1819), vol. 1, p. 187.
[31] David Hume, 'Of the Rise and Progress of the Arts and Sciences,' cited in Strawson, *Real Materialism*, p. 423, n. 25.
[32] 'The Stony Brooks Interviews, Interview 4: Chomsky on the Mind,' interviewed by Peter Ludlow, in Peter Ludlow, *The Philosophy of Generative Linguistics* (Oxford University Press, 2011), p. 174.
[33] Noam Chomsky, 'The Machine, the ghost, and the limits of understanding: Newton's contributions to the study of mind,' The CSMN Annual Lecture on Mind in Nature, University of Oslo, September 2011: http://www.youtube.com/watch?v=D5in5EdjhD0.
[34] Chomsky, *New Horizons in the Study of Language and Mind*, p. 62.
[35] Bertrand Russell, *Human Knowledge: Its Scope and Limits* (New York: Simon & Schuster, 1948).
[36] Russell, *The Problems of Philosophy*, p. 42 (emphasis his).
[37] Immanuel Kant, *Critique of Pure Reason*, trans. Norman Kemp Smith (London: St Martin's Press, 1963), p. 183.
[38] Wilhelm von Humboldt, *On Language: On the Diversity of Human Language-Structure and its Influence on the Mental Development of Mankind*, trans. Peter Heath (Cambridge University Press, 1988), p. 59 (emphasis his).
[39] Cited in Alain de Botton, 'Montaigne rescue,' review of James Miller, *The Philosophical Life: 12 Great Thinkers and the Search for Wisdom, from Socrates to Nietzsche*, *New Statesman*, 20 February 2012.
[40] Chomsky, *New Horizons in the Study of Language and Mind*, p. 4.
[41] Ibid., p. 118.
[42] Cited in Ottor Jespersen, *The Philosophy of Grammar* (London: George Allen & Unwin, 1924), p. 47.
[43] Noam Chomsky, 'Recent contributions to the theory of innate ideas,' *Synthese*, 17(1), March 1967 (Dordrecht, Holland: D. Reidel Publishing Company): 3-4.
[44] H. Clark Barrett, 'Where there is an adaptation, there is a domain: The form-function fit in information processing,' *Foundations in Evolutionary Cognitive Neuroscience*, eds.

Steven M. Platek and Todd K. Shackelford (Oxford University Press, 2009), p. 112 (97-116).
[45] Jespersen, *The Philosophy of Grammar*, p. 19.
[46] Chomsky, 'Recent contributions to the theory of innate ideas,' 8.
[47] Richard Lewontin, *The Doctrine of DNA: Biology as Ideology* (London: Penguin, 1993), pp. 36-7.
[48] Noam Chomsky, *The Minimalist Program* (MIT Press, 1995), p. 8.
[49] Noam Chomsky, 'A Minimalist Program for Linguistic Theory,' in *The View from Building 20: Essays in Linguistics in Honor of Sylvain Bromberger,* ed. Kenneth Hale and Samuel Jay Keyser, (MIT Press, 1993, repr. 1999), p. 44.
[50] 'Rationalism vs. Empiricism,' Stanford Encyclopedia of Philosophy, http://plato.stanford.edu/entries/rationalism-empiricism.
[51] Chomsky, *New Horizons in the Study of Language and Mind*, p. 122.
[52] Noam Chomsky, *The Science of Language: Interviews with James McGilvray* (Cambridge University Press, 2012), p. 52.
[53] Hinzen, *Mind Design and Minimal Syntax*, p. 124, n. 4.
[54] S. J. Hespos and E. S. Spelke, 'Conceptual precursors to language,' *Nature*, 430, 2004: 453-6. See also E. S. Spelke, 'The origins of physical knowledge,' *Thought without Language*, ed. L. Weiskrantz (Oxford University Press, 1988), pp. 168-84.
[55] Paul Bloom, *How Children Learn the Meanings of Words* (MIT Press, 2000), p. 262.
[56] M. C. Anderson, K. N. Ochsner, B. Kuhl, J. Cooper, E. Robertson, S. W. Gabrieli, G. H. Glover, J. D. E. Gabrieli, 'Neural Systems Underlying the Suppression of Unwanted Memories,' *Science*, 303(5655), 9 January 2004: 232-235.
[57] Gennaro Chierchia, 'Language, Thought and Reality after Chomsky,' *Chomsky Notebook*, ed. Jean Bricmont and Julie Franck (Columbia University Press, 2010), pp. 166-7.
[58] Norman Segalowitz, 'On the evolving connections between psychology and linguistics,' *Annual Review of Applied Linguistics*, vol. 21 (2001): 9-10; reviewing M. I. Posner and S. E. Peterson, 'The attention system of the human brain,' *Annual Review of Neuroscience*, vol. 13 (1990): 25-42.
[59] Ray Jackendoff, *Foundations of Language: Brain, Meaning, Grammar, Evolution* (Oxford University Press, 2002), p. 22.
[60] Ibid., p. 58.
[61] Pierre Jacob and Marc Jeannerod, *Ways of Seeing: The Scope and Limits of Visual Cognition* (Oxford University Press, 2003), p. 17.
[62] D. M. Rosenthal, 'Two concepts of consciousness,' *Philosophical Studies*, 99, 1986: 329-59.
[63] V. A. Fromin, 'Some thoughts about the brain/mind/language interface,' *Lingua*, 100, 1997: 23 (3-27).
[64] Don Garrett, *Cognition and Commitment in Hume's Philosophy* (Oxford University Press, 1997), p. 171.
[65] Noam Chomsky, *Problems of Knowledge and Freedom* (London: Barrie and Jenkins, 1972), p. 41.
[66] Antonio Damasio, *Looking for Spinoza: Joy, Sorrow, and the Feeling Brain* (Orlando, FL.: Harcourt, 2003), p. 54.
[67] Iain McGilchrist, *The Master and His Emissary: The Divided Brain and the Making of the Western World* (Yale University Press, 2010), p. 119. See for example D. Vowles, 'Neuroethology, evolution and grammar,' in R. Aronson, E. Tobach and D. Lehrman et al (eds.), *Development and Evolution of Behavior: Essays in Memory of TC Schneirla* (San Francisco: W. H. Freeman & Co., 1970): 194-215.
[68] Ibid., p. 80. See A. Carstairs-McCarthy, *The Origins of Complex Language: An Inquiry into the Evolutionary Beginning of Sentences, Syllables, and Truth* (Oxford University Press, 1999).

[69] McGilchrist, *The Master and His Emissary*, p. 123.
[70] George Lakoff and Mark Johnson, *Philosophy in the Flesh: The Embodied Mind and its Challenge to Western Thought* (New York: Basic Books, 1999), p. 4.
[71] Cited in Noam Chomsky, *Knowledge of Language: Its Nature, Origin, and Use* (Westport, CT: Greenwood Publishing Group, Inc., 1986), p. xiv.
[72] Marc. D. Hauser, Noam Chomsky, W. Tecumseh Fitch, 'The Faculty of Language: What Is It, Who Has It, and How Did It Evolve?', *Science*, 298(5598), 2002: 1570.
[73] Ibid.
[74] Hinzen, *Mind Design and Minimal Syntax*, p. 130, n. 6.
[75] For further discussion see *The View from Building 20: Essays in Linguistics in Honor of Sylvain Bromberger*, ed. Kenneth Hale, Samuel Jay Keyser, (MIT Press, 1993, repr. 1999).
[76] John A. Lucy, *Language Diversity and Thought: A Reformulation of the Linguistic Relativity Hypothesis* (Cambridge University Press, 1996 [1992]), p. 14.
[77] John Locke, in James W. Underhill, *Humboldt, Worldview and Language* (Edinburgh University Press, 2009), p. 64.
[78] Hinzen, *Mind Design and Minimal Syntax*, p. 19, n. 17
[79] Ray Jackendoff, *A User's Guide to Thought and Meaning* (Oxford University Press, 2012), p. 15.
[80] Gottlieb Frege, 'On Sense and Reference,' trans. Max Black, *Meaning and Reference*, ed. A. W. Moore (Oxford University Press, 1993), p. 26.
[81] Gottlob Frege, 'The thought,' *Mind*, 65(259), 1956: 292 (289-311).
[82] Jackendoff, *A User's Guide to Thought and Meaning*, p. 15, n. 2.
[83] Charles Morris, 'Foundations of the Theory of Signs,' *Writings on the General Theory of Signs* (The Hague: Mouton, 1971), p. 28, 35, 43.
[84] John Searle, 'Philosophy 132,' Lecture 001, Spring 2012, UC Berkeley, released 27 April 2012 on iTunes U.
[85] Strawson, *Real Materialism and Other Essays*, p. 8.
[86] Hinzen, *Mind Design and Minimal Syntax*, p. 18.
[87] Alonzo Church, *The Calculi of Lambda Conversion* (Princeton University Press, 1941).
[88] Hinzen, *Mind Design and Minimal Syntax*, p. 72.
[89] Zeno Pylyshyn, *Seeing and Visualizing: It's Not What You Think* (Massachusetts: MIT Press, 2003), p. 38.
[90] Ray Jackendoff, *Language, Consciousness, Culture: Essays on Mental Structure* (Massachusetts: MIT Press, 2007), p. 106.
[91] Richard Kayne, *Parameters and Universals* (Oxford University Press, 2000), p. 7.
[92] Norbert Hornstein and Louise Anthony, 'Introduction,' *Chomsky and His Critics*, eds. Louise M. Antony and Norbert Hornstein (Oxford: Blackwell, 2003), p. 9.
[93] Cited in Chomsky, *New Horizons in the Study of Language and Mind*, p. 136.
[94] Jackendoff, *Foundations of Language*, p. 35.
[95] Robert C. Berwick, Paul Pietroski, Beracah Yankama and Noam Chomsky, 'Poverty of the Stimulus Revisited,' *Cognitive Science*, 35(7), September/October 2011: 1207 (1207-42).
[96] Paul Horwich, 'Meaning and its Place in the Language Faculty,' *Chomsky and His Critics*, p. 165.
[97] Hauser, Chomsky and Fitch, 'The Faculty of Language,' 1571.
[98] Chomsky, 'Recent contributions to the theory of innate ideas,' 8.
[99] Ibid; Wilhelm von Humboldt, *On Language: the Diversity of Human Language Construction and its Influence on the Mental Development of the Human Species*, 2nd ed., ed. M. Lomansky, trans. P. L. Heath (Cambridge University Press, 1999), p. 122.
[100] Colin McGinn, 'Imagination,' *The Oxford Handbook of Philosophy of Mind*, ed. Brian P. McLaughlin, Ansgar Beckermann and Sven Walter (Oxford University Press, 2009), p. 604 (emphasis his).

[101] *A Bit of Fry & Laurie*, BBC, 1989, Series 1, Episode 2.
[102] Hauser, Chomsky and Fitch, 'The Faculty of Language,' 1578.
[103] Underhill, *Humboldt, Worldview and Language*, p. 58.
[104] Dennis Hassabis, 'Model the brain's algorithms,' *Nature*, 482(7386), 23 February 2012: 462.
[105] David Hume, *An Enquiry Concerning Human Understanding*, in *Enquiries Concerning the Human Understanding and Concerning the Principles of Morals*, ed. L. A. Selby-Bigge, 2nd ed. (Oxford: Clarendon Press, 1902), Section IX, 'Of the Reason of Animals,' p. 108.
[106] Chomsky, *New Horizons in the Study of Language and Mind*, p. 8.
[107] Jackendoff, *Foundations of Language*, p. 102.
[108] Aristotle, 'On the Art of Poetry,' in *Classical Literary Criticism*, trans. T. R. Dorsch (Harmondsworth: Penguin, 1965), p. 35.
[109] Cedric Boeckx, 'Some Reflections on Darwin's Problem in the Context of Cartesian Biolinguistics,' *The Biolinguistic Enterprise: New Perspectives on the Evolution and Nature of the Human Language Faculty*, ed. Anna Maria di Sciullo and Cedric Boeckx (Oxford University Press, 2011), p. 62.
[110] A. N. Meltzoff and M. K. Moore, 'Explaining facial imitation. A theoretical model,' *Early Development and Parenting*, 6, 1997: 179-92; A. N. Meltzoff and M. K. Moore, 'Imitation of facial and manual gestures by human neonates,' *Science*, 7 October 1977, 198(4313): 75-8.
[111] Chomsky, 'Recent contributions to the theory of innate ideas,' 9.
[112] Arthur Koestler, *The Ghost in the Machine* (London: Hutchinson, 1967), p. 19.
[113] John Napier, *Hands* (London: George Allen Unwin, 1980), p. 116.
[114] Edward Sapir, 'The unconscious patterning of behavior in society,' in E. S. Dummer (ed.) *The Unconscious: A Symposium* (New York: Knopf, 1927), pp. 114-42.
[115] McGilchrist, *The Master and His Emissary*, p. 71.
[116] V. S. Rotenberg and V. V. Arshavsky, 'The two hemispheres and the problem of psychotherapy, *Dynamische Psychiatrie/Dynamic Psychiatry*, 20 (5-6), 1987: 371.
[117] Nathanaël, *from* THE MIDDLE NOTEBOOKES [sic], *Chicago Review*, 56(2), Autumn 2011: 117, 119, 121.
[118] Friedrich Albert Lange, *The History of Materialism and Criticism of its Present Importance*, trans. Ernest Chester Thomas, 3rd ed. (London: Routledge & Kegan Paul, 1957), p. 77.
[119] Charles Dickens, *A Tale of Two Cities*, I, iii.
[120] Isaiah Berlin, *The Roots of Romanticism* (Princeton University Press, 1999), pp. 102-4.
[121] Elizabeth Spelke, 'Innateness, Choice, and Language,' in *Chomsky Notebook*, p. 208.
[122] Donald Davidson, 'The structure and content of truth,' *Journal of Philosophy*, 87(6), 1990: 314 (279-328).
[123] Noam Chomsky, *Rules and Representations* (Columbia University Press, 1980), p. 229-30.
[124] Russell, *The Problems of Philosophy*, p. 32.
[125] Hinzen, *Mind Design and Minimal Syntax*, p. xii.
[126] Cited in Gillian Rudd, *Managing Language in Piers Plowman* (Cambridge: D. S. Brewer, 1994), p. 197.
[127] James Higginbotham, 'On Semantics,' *Linguistic Inquiry*, 16(4), Fall 1985: 547-93.
[128] Berwick, Pietroski, Yankama and Chomsky, 'Poverty of the Stimulus Revisited,' 1213.
[129] Ibid, 1218.
[130] Robert C. Berwick and Noam Chomsky, 'The Biolinguistic Program: The Current State of its Development,' *The Biolinguistic Enterprise*, p. 25.
[131] John Searle, *The Construction of Social Reality* (New York: Free Press, 1995).

[132] Humboldt, *On Language*, p. 159.
[133] Cited in Noam Chomsky, *On Nature and Language* (Cambridge University Press, 2002), pp. 76-7.
[134] Cited in Daniel Dennett, *Consciousness Explained* (London: Penguin, 1993), p. 194.
[135] Tony Stone and Martin Davies, 'Chomsky Amongst the Philosophers,' review of Noam Chomsky, *New Horizons in the Study of Language and Mind*, *Mind & Language*, vol. 17, no. 3 (June 2002): 286.
[136] Humboldt, *On Language*, p. 220.
[137] Noam Chomsky, 'Three Factors in Language Design,' *Linguistic Inquiry*, 36(1), Winter 2005: 3-4; citing Francois Jacob, *The Possible and the Actual* (New York: Pantheon, 1982), p. 59.
[138] Chomsky, *New Horizons in the Study of Language and Mind*, p. 71.
[139] Hinzen, *Mind Design and Minimal Syntax*, p. 90.
[140] Ibid., p. 128.
[141] Gordon Claridge, 'Schizophrenia and Human Individuality,' *Mindwaves: Thoughts on Intelligence, Identity and Consciousness*, ed. Colin Blakemore and Susan Greenfield (Oxford: Basil Blackwell, 1987), p. 31 (29-41).
[142] Robert C. Berwick and Noam Chomsky, 'The Biolinguistic Program: The Current State of its Development,' *The Biolinguistic Enterprise*, p. 32.
[143] Chomsky, *New Horizons in the Study of Language and Mind*, p. 122.
[144] Robert C. Berwick and Noam Chomsky, 'The Biolinguistic Program: The Current State of its Development,' *The Biolinguistic Enterprise*, p. 26, quoting Ian Tattersall, *The Origin of the Human Capacity*, series James Arthur Lecture on the Evolution of the Human Brain 68 (New York: American Museum of Natural History, 1998).
[145] Robert C. Berwick and Noam Chomsky, 'The Biolinguistic Program: The Current State of its Development,' *The Biolinguistic Enterprise*, p. 33.
[146] A. Benítez-Burraco, *Genes del Lenguaje: Implicaciones Ontogenéticas, Filogenéticas y Cognitivas*, PhD dissertation, University of Oviedo, 2007.
[147] W. Tecumseh Fitch, '"Deep Homology" in the Biology and Evolution of Language,' *The Biolinguistic Enterprise*, p. 163.
[148] Karl C. Diller and Rebecca L. Cann, 'Genetic Influences on Language Evolution: An Evaluation of the Evidence,' *The Oxford handbook of Language Evolution*, ed. Maggie Tallerman and Kathleen R. Gibson (Oxford University Press, 2012), pp. 174-5 (pp. 168-75).
[149] Patricia Kuhl and James Miller, 'Speech perception by the chinchilla: Identification function for synthetic VOT stimuli,' *Journal of the Acoustical Society of America*, 63, 1978: 905-17.
[150] McGilchrist, *The Master and His Emissary*, p. 106.
[151] J. M. Coetzee, *Disgrace* (London: Vintage, 2000), p. 4.

2

LAWS OF FORM

Therefore this terror and darkness of the mind
Not by the sun's rays, nor the bright shafts of day,
Must be dispersed, as is most necessary,
But by the fact of nature and her laws.

LUCRETIUS[1]

Adopting Husserl and Weinberg's 'Galilean style' of naturalism for the study of language, the minimalist program (or standard scientific practice in a different name) assumes with the physicists that theory-formation 'proceed[s] independently of the world and experience; the actual theory or model one ends up with is an expression of the creative activity of the mind, modulated by what experience suggests.'[2] Julian Huxley seems to have touched upon something similar in 1926:

> The man who is content to make records or to collect skins and eggs will, unless he spends years of his life in a systematic analysis of his own and others' facts, not get anything from his labours – save the very real pleasure of making the observations. But he who takes the trouble to think out new problems and new lines of attack upon the old will have the same pleasure, and in addition the joy of intellectual discovery.[3]

The biolinguistic stance towards the 'mental' aspects of the world has generally been that

> The role of experiment is to *demonstrate existence* for the *nature* that one has determined. It creates an *effect*, for a known *cause*, in much the Cartesian sense. It is not that the cause *becomes known* to us through the experiment: this would be like saying that by seeing a concept that figures in our understanding to be instantiated, we thereby know the concept. On the contrary, we have to know the concept in order to see *it* instantiated.[4]

Seeking what some regard as perfection in nature has been 'a very profitable objective,' wrote Dirac.[5] 'Perfection' here should be understood as being more of an aesthetic than functional notion, as when Antoine de Saint-Exupéry in *The Little Prince* mused that 'whether it's a house, or the stars, or the desert, what makes them beautiful is invisible.' Though 'perfection' is to be expected in the physical and not organic world, 'perfection' in the minimalist sense of satisfying specifications for minimal design may be found in FLN. Some have consequently claimed 'linguistics and physics have the same Galilean character.'[6] From this perspective, minimalism is simply Galilean science applied to the mind. 'Minimalism' and 'internalism' (which stress the importance of structural complexity over external shaping effects) is also the standard position in physics: the solar system has no 'function,' but it does have an internal structure, dictated by natural law.

Even in 1968 it was recognised by Robert Sklar that linguistics 'is really theoretical biology.'[7] A few years earlier, Turing's main work on morphogenesis showed that 'if you ever managed to understand anything really critical about biology, you'd belong to the chemistry or physics departments.'[8] Following the spirit of Occam and Galileo, Chomsky has suggested that UG incorporates a 'least effort condition,' or the Economy Principle; namely, that 'Derivations and representations ... are required to be minimal ... with no superfluous steps in derivations and no superfluous symbols in representations.'[9] These assumptions embrace the possibility that an 'inherently global principle of computational optimization further forces syntactic derivation to adopt some sort of computational cycles, such as *phases*, constituting a kind of heuristic "computational trick" that syntax uses for restricting computational domains locally and thus reducing the computational load.'[10] Along with Eric Lenneberg's largely forgotten 1967 *Biological Foundations of Language*, Chomsky's 1965 *Aspects of the Theory of Syntax* also hinted at a minimalist-style approach to language, noting the importance of 'principles of neural organization that may be even more deeply grounded in physical law.'[11] The minimalist/biolinguist, then, is simply investigating a natural system and its satisfaction of interface conditions, ignoring questions of ontology or disciplinary labels of 'linguistics' or 'psychology.' The assumption for the physicist and the biolinguist is that nature realises 'perfect' solutions (hence the term 'minimalism'), and only if the assumption turns out to be false (as is almost always the case in biological organs) then the theory adjusts to the requirements imposed by the empirical data. This minimalist program for theoretical linguistics is largely motivated by 'the desire to minimize the acquisition burden on the child, and thereby maximize the learnability of natural language grammars.'[12] Minimalism's search for elementary principles of cognition and computation (like Merge) makes it a 'fundamental' theory (which outlines the metaphysical basis) rather than an 'effective' one (which roughly matches empirical data), to borrow the terms of the physicists. The above minimalist concerns are also characteristic of the sciences more generally:

> Now in the further advancement of science, we want more than just a formula. First we have an observation, then we have numbers that we measure, then we have a law which summarizes all the numbers. But the real glory of science is that we can find a way of thinking such that the law is evident.[13]

This naturalistic framework has its roots in classical thought, with Olympiodorus for instance claiming that 'Nature does nothing superfluous or any unnecessary work.'[14] Formulating his first rule of reasoning in 'natural philosophy,' Newton wrote that 'nature is pleased with simplicity, and affects not the pomp or superfluous causes.'[15] Bacon's words compliment a similar approach, and seem far more relevant today than for the mechanical philosophers: 'The subtlety of Nature is greater many times over than the subtlety of the senses and understanding.'[16] A search for perfection and beauty was something which also guided Copernicus, who began *The Revolutions of the Heavenly Spheres* by confessing that 'among the many and varied literary and artistic studies upon which the natural talents of man are nourished, I think that those above all should be embraced and pursued with the most loving care which have to do with things that are very beautiful.'[17] A similar recognition of the limits of cognition can also be found during his exposition against the Earth-centric theories of Ptolemy:

> Why therefore should we hesitate any longer to grant it the movement which accords naturally with its form, rather than put the whole world in a commotion – the world whose limits we do not and cannot know? And why not admit that the appearance of daily revolution belongs to the heavens but the reality belongs to the Earth?[18]

We can sympathise with Abelard, then, when he announced that 'The man of understanding is he who has the ability to grasp and ponder the hidden causes of things. By hidden causes we mean those from which things originate, and these are to be investigated more by reason than by sensory experience.'[19] Rousseau also observed that 'Nature never deceives us; it is always we who deceive ourselves.'[20] But if Abelard and Rousseau are correct, then we must recognise with Russell that we are each comprised of 'a body whose sense-organs distort as much as they reveal.'[21] Science, then, is based on explanatory conveniences, while our perception of the world is based on sensory conveniences. To take just one example, it was thanks to Poincaré's work on geometry that 'the choice if a curved or a flat space-time in [the theory of general relativity] was, once again, simply a conventional choice of "convenience" on our part.'[22] The Australian philosopher Frank Jackson takes this further: 'It is a commonplace that there is an apparent clash between the picture Science gives of the world around us and the picture our senses give us. We *sense* the world as made up of coloured, materially continuous, macroscopic, stable objects; Science and, in particular, Physics, tells us that the material world is constituted of clouds of minute, colourless, highly-mobile particles.'[23]

In any domain of naturalistic inquiry, 'Questions arise when we find things not to be as we think they should be,' as when the planets were discovered to be moving in ellipses and not perfect spheres.[24] The assumptions have appeared suitable in physics, and there is little reason to believe they are unfit for inquiry into the 'mental' world, which has presumably been fashioned by the same natural laws and physical constraints which govern the 'chemical' and 'organic' worlds, miracles aside. One striking outcome of this minimalist approach for semantics (which will, if true, no doubt have its repercussions felt in philosophy departments around the country in a few generation, with mainstream philosophical dogmas taking their sweet time disappearing) is that it entails, as Paul Pietroski believes, the possibility that meaning is purely a by-product of structural growth and an emergent product of lexical interactions during syntactic derivation.

As mentioned, functional perfection is not to be expected in the organic world, unlike in the physical universe, since nature is 'an opportunistic maker of gadgets, a "satisficer" who is always ready to settle for mediocrity if it is cheap enough.'[25] Like other biological systems, language is imperfect, with the human cognitive system as a whole being 'a kludge, a collection of *ad hoc* systems that somehow get the job done.'[26] This is possibly one of the main reasons the Galilean style has often been regarded as a weakness in biology, to be regarded as a lack of 'satisfying explanation' requiring 'more measurement and less theory,' as Fox-Keller notes of mainstream biology.[27] Lacking such prejudice, biolinguistics stresses the need for addressing the evolution of language 'from a multi-disciplinary perspective, to avoid the facile adaptationist, just-so story traps that are all too familiar' – what William Blake would have called '*the* Science, of Imaginary

Solutions,' such as Pinker and Bloom's applauded 1990 essay 'Natural Language and Natural Selection.'[28] The Neo-Darwinian Synthesis is fine for explaining an organism's survival, but says absolutely nothing about why it took the form and structure it did to begin with. An archetypal functionalist and adaptationist view was presented by Wartofsky's standard introduction to the philosophy of science:

> Following the lead of the evolutionary biologist, we may consider the broad thesis that knowing activities ... have evolved from precognitive responses and adaptation to an environment and that they are the product of natural and cultural selection, in the sense that they have survival value for the human species.[29]

Much of modern evolutionary theory has moved beyond this 'micro mutational' gradualist view, symbolised by the 'hill-climbing' metaphor William James derided as 'Pop-Darwinism' in his critique of Herbert Spencer. Natural selection, Dawkins assures us, is 'a relentless slight bias in death rates, or a slight bias in reproductive rates, among the individuals within a species. Given enough time, small changes in lineages diverge yet further until we have descendents as different as elephants from moles, or squids from warthogs.'[30] He points out that 'A gene is a sequence of code letters, drawn from an alphabet of precisely four letters, and the genetic code is universal throughout all known living things. Life is the execution of programs written using a small digital alphabet in a single, universal machine language.'[31] This gene-centric Neo-Darwinian Synthesis has been increasingly undermined by the 'developmentalist challenge,' since much more is needed to explain organic nature than genes and environment alone.[32] If certain of these thinkers have been accused of 'physics envy,' Dawkins is certainly guilty of 'poetry envy,' strengthening the imagery of Pop-Darwinism with his book *Climbing Mount Improbable* and similar odes. Internalist accounts view 'adaptation as secondary tinkering rather than primary structuring.'[33] In evolutionary psychology, 'Assuming a priori that a trait may be an adaptation is an experimental heuristic that guides research questions and methodology,' putting aside concerns about internal structure and laws of form.[34] What are often viewed as adaptive problems are also not necessarily so, but rather the 'regularities of the physical, chemical, developmental, ecological, demographic, social, and informational environments encountered by ancestral populations during the course of a species' or population's evolution.'[35] The adaptationists even seem to hearken back (perhaps unintentionally) to August Weismann's argument for the *Allmacht*, or omnipotence, of natural selection in evolutionary history.[36] We should also be quick, however, not to cast aside speculations on the changing roles of the various mechanisms involved in externalization, like the view that 'the development of manufacturing and tool use have selective advantage to shifting the burden of communication from the hands to the face,' as Michael Corballis observes.[37]

Nevertheless, evolutionary psychology and much of modern cognitive science is grounded in the ultra-Neo Darwinian perspective of the mind as a product of adaptive pressures. Language is therefore expected to be constructed from an array of 'loosely interacting computational tricks.'[38] If that were true, writes Cedric Boeckx, 'the search for a restricted core of deep abstract principles of optimal design would be doomed from the start. Fortunately for minimalists, the view of nature as a tinkerer is being re-evaluated in biology, and the search for deep organizational motifs is slowly coming to fruition.' Following the model of theoretical biology constructed by D'Arcy Thompson along with the emerging evo-devo program, 'minimalism may well turn out to provide remarkable support for a silent revolution in biology.'[39] A recent paper on the physical genesis of multicellular forms also concludes that, 'rather than being the result of evolutionary adaptation, much morphological plasticity reflects the influence of external physico-chemical parameters on any material system and is therefore an inherent, inevitable property of organisms.'[40]

Hinzen's position, and that of biolinguistics more generally, is 'essentially parallel to one found in theoretical biology, where a position that its nineteenth century defenders called "formalism" or "rational morphology" allowed for the autonomous study of animal form, disregarding the external *conditions of existence* that drive such organic forms in or out of existence on the evolutionary scene.'[41] Whereas the functionalist asks what certain capacities are *for*, the formalist will stress the importance of principles determining organic structural complexity. Whereas many have characterised the central debate in nineteenth-century biology as being between evolutionists and creationists, a more accurate classification (as Darwin himself noted) would distinguish teleologists (who regarded adaptation and the single most important aspect in evolution) with morphologists (who held that commonalities of structure were the defining biological characteristic) – a dichotomy stressed by E. S. Russell's *Form and Function*.[42] The formalists (or, following Chomsky's terminology, 'internalists') include such figures as D'Arcy Thompson, Brian Goodwin, Richard Owen, Stuart Kauffman, Geoffroy St. Hilaire, Richard Goldschmidt, Nikolai Severtzov, Louis Agassiz, Karl Ernst von Baer and Goethe (whose plant studies lead him to coin the term 'rational morphology'). They focused on form and structural commonalities as their explanadum, leaving aside the question of adaptive effects as a secondary concern.[43] Richard Owen, for instance, stressed what he called 'Unity of Type' over 'conditions of existence' – concerns which theoretical biologists like Thompson would later emphasize. Darwin's rival Wallace also objected to the claim that natural selection alone could explain human arithmetic capacities, scarcely employed throughout history and largely hidden in the head. Darwin himself proposed in *The Descent of Man* that the only distinction to be made between humans and other animals was that man differs 'solely in his almost infinitely larger power of associating together the most diversified sounds and ideas; and this obviously depends on the high development of his mental powers.'[44] He was wrong to assert 'solely' and 'infinitely' rather than 'unbounded' (as in discrete infinity), but the importance he draws to language as a defining characteristic of humans has not diminished since 1871. In 1949 one of the most influential palaeontologists of the twentieth century, G. G. Simpson, reinforced this Darwinian (and Lockean) perspective:

> Man arose as a result of the operation of organic evolution and his being and activities are also materialistic, but the human species has properties unique to itself among all form of life, superadded to the properties unique to life among all forms of matter and of action. Man's intellectual, social, and spiritual natures are altogether exceptional among animals in degree, but they arose by organic evolution.[45]

But to adopt the essence of Wallace's point (despite its theological underpinnings), 'evolution' involves many factors, not just natural selection. These internalist concerns are also found in the perceptive words of the seventeenth-century German physician Daniel Sennert:

> Always in vain does anyone resort to external causes for the concreation of things; rather does it concern the internal disposition of the matter. ... On account of their forms ... things have dispositions to act. ... They receive their perfection from their form, not from an external cause. Hence also salt has a natural concreation ... not from heat or cold, but from its form, which is the architecture of its domicile.[46]

Reviving these largely forgotten considerations, biolinguistic internalism is opposed to a Lamarckian interpretation of evolution in which external factors determine internal structure, with internal structure satisfying these external demands. Adaptationism is not to be discouraged altogether, and can contribute to evolutionary theory so long as it arises 'where laws of form have paved a way' (making the adaptationism/formalism debate similar to the empiricism/rationalism debate of the eighteenth and nineteenth centuries).[47] Though not a formalist himself, L. T. Hobhouse (early opposer of the Boer War and Britain's innovative use of concentration camps in South Africa) stressed in his 1901 study *Mind in Evolution* that the chaotic motion both of long grass and of 'the white blood-corpuscle' are 'only very complicated results of the same set of physical laws in accordance with which the grass bows before the wind.'[48]

Criticising the functionalists and adaptationists (such as Dawkins, Fisher, Dobzhansky, Jackendoff and Pinker), Hinzen notes that the existence of something does not necessarily concern the *nature* of that thing, a notion which 'surfaces in Sartre's famous existentialist argument that there is no such thing as human nature, for from mere *existence* (Heidegger's *Geworfenheit*), no human *nature* could be extracted. The existentialist's point is that the conditions under which a thing exists do not only leave its nature *underspecified*; as such they do not determine it *at all*.'[49] Hinzen's claims are similar to the Kantian notion of existence not being a predicate or a property of mind-external 'things' but of 'concepts.' These are also some of the reasons why the term 'evolution' has come to be more than misleading at times, since the concept itself does not stipulate precisely what factors enter into an organism's evolution (which include physical constraints, environmental and social pressures, and so on). Adapting to an environment to increase the chances of survival and gene propagation is commendable, but the internalist would stress that this process takes place within the context of physical constraints and a set of options conferred by what Owen called an 'archetype,' or an 'ideal type that embodied the potential for the all the fundamental structures of the body and limited the variation possible within members of a single group of animals.'[50]

The goals of Theoretical Morphology, outlined by George McGhee, are probably the closest to those of biolinguistics:

> The goal is to explore the possible range of morphologic variability that nature could produce by constructing n-dimensional geometric hyperspaces (termed "theoretical morphospaces"), which can be produced by systematically varying the parameter values of a geometric model of form. ... Once constructed, the range of existent variability in form may be examined in this hypothetical morphospace, both to quantify the range of existent form and to reveal nonexistent organic form. That is, to reveal morphologies that theoretically could exist ... but that never have been produced in the process of organic evolution on the planet Earth. The ultimate goal of this area of research is to understand why existent form actually exists and why nonexistent form does not.[51]

Opposing the biolinguists, Jackendoff points out that 'Chomsky is correct that other factors besides natural selection play a role in evolution, for instance structural constraints (no terrestrial animal will develop wheels) and the biochemistry of proteins. Nevertheless, there is nothing in these other factors (so far) that provides any helpful hints on what caused language to emerge.'[52] Be at as it may, the stress of Lewontin and Levins on 'third factor' effects has not faded in relevance since 1985: 'Natural selection is not a consequence of how well the organism solves a set of fixed problems posed by the environment; on the contrary, the environment and the organism actively codetermine each other. The internal and the external factors, genes and environment, act upon each other through the medium of the organism.'[53] To understand organic development, write West and King, we should 'ask not what's inside the genes you inherited, but what the genes you inherited are inside of.'[54] Similar priorities have lead James McGilvray has suggested that instead of using Massimo Piatelli-Palmarini's term 'biolinguistics' (introduced in 1974 as the title of a linguistics conference held in Royaumont, France), 'perhaps we should speak instead of "biophysical linguistics" or perhaps "bio-compuphysical linguistics," so that it becomes clear that the set of possible natural languages and I-languages depends not just on genetic coding but also on other factors – all, though, conceived of as somehow built into nature and the ways in which it permits development/growth.'[55] 'Third factors' can also be aptly characterised by Kurt Lewin's notion of Galilean explanations, as opposed to Aristotelian ones, a distinction in the physical sciences which Boeckx has recently carried over to the cognitive sciences:

> Aristotelian laws or explanation have the following characteristics: they are *recurrent*, that is statistically significant; they specifically (though not always) target functions, that is they have a *functionalist* flavor to them; they also *allow for exceptions*, organized exceptions or not, but at least they allow for exceptions; and finally they have to do with *observables* of various kinds. ... [Galilean laws] are typically *formal* in character, and they are very abstract mathematically. They *allow for no exceptions* and they are *hidden*. That is, if you fail to find overtly the manifestation of a particular law that you happen to study, this does not mean that it is not universal. It just means that it is hidden and that we have to look at it more closely and we will eventually see that the law actually applies.[56]

Galileo also famously used his stress on idealisations to combat religious dogmatism, as when he wrote in his 1615 'Letter to the Grand

Duchess Christina of Lorraine': 'I do not feel obliged to believe that that same God who has endowed us with senses, reason, and intellect has intended to forgo their use and by some other means to give us knowledge which we can attain by them. He would not require us to deny sense and reason in physical matters which are set before our eyes and minds by direct experience or necessary demonstrations.'[57] Commenting on the philosophy of the Italian physicist, Leonardo Olschki wrote in a 1943 edition of *Philosophical Review* that 'no mediator is needed for our cognition of the divine intelligence as soon as we possess the tools supplied by this progressive science. For Galileo metaphysics does not lie beyond physics because physics is a part of metaphysics. Thus theology and natural science have, though their fields and methods are different, an identical goal.'[58]

Unlike the Neo-Darwinian focus on adaptation, the developmental biologists seek to explain the origin of internal structure and organic form, often constrained by Lewin's hidden, Galilean laws. In the early nineteenth century, Geoffroy St. Hilaire voiced similar concerns when he argued that formalism could be summed up with the phrase 'such is the organ, such will be its function.'[59] Ernst Mayr, in his deeply-informed study *What Makes Biology Unique?*, wrote that in the study of the organic world 'to have isolated all parts, even the smallest ones, is not enough for a complete explanation of most systems,' noting further that 'laws certainly play a rather small role in theory construction in biology' 'because evolutionary regularities do not deal with the basics of matter as do the laws of physics. They are invariably restricted in space and time, and they usually have numerous exceptions.'[60] Evo-devo is the attempt to unify biology through the integration of evolutionary and developmental biology, with geneticist Theodosius Dobzhansky's Neo-Darwinian Synthesis being contradicted by recent research: 'Evolution, it turns out, makes no sense except in the light of biology – developmental biology, to be precise.'[61]

But Dobzhansky's thinking also retaliated against the then popular view of the human mind being a 'blank slate.' The modular brain, on the other hand, if considered part of the human body and not of the aether, 'is no more plausibly a blank slate with unlimited plasticity in response to its environment than is the mind.'[62] Darwin's bulldog, Thomas Huxley, also explained that, 'in view of the intimate relations between Man and the rest of the living world, and between the forces exerted on the latter and all other forces, I can see no excuse for doubting that all are co-ordinated terms of Nature's great progression, from the formless to the formed – from the inorganic to the organic – from blind force to conscious intellect and will.'[63] In his 1962 *Mankind Evolving*, Dobzhansky wrote: 'A newborn infant is not a blank slate;

however, his genes do not seal his fate. His reactions to the world around him will differ in many ways from those of other infants, including his brothers. My genes have indeed determined what I am, but only in the sense that, given the succession of environments and experiences that were mine, a carrier of a different set of genes might have become unlike myself.'[64] This genetic revolution signalled, for good, the end of dualistic excuses for souls and other immaterial forces at work 'in' the body. Though this doesn't stop our dualist intuitions pervading our concepts of insensate objects versus animate bodies: We cannot, for instance, say 'Sam touched the library on the window,' since the library doesn't *have* a window, it simply *is* the window (more on this later).[65]

Like the Scottish intellectuals of the Enlightenment, Russell understood that 'All knowledge, we find, must be built up upon our instinctive beliefs, and if these are rejected, nothing is left. But among our instinctive beliefs some are much stronger than others, while many have, by habit and association, become entangled with other beliefs, not really instinctive, but falsely supposed to be part of what is believed instinctively.'[66] This illustrates all too clearly how the remnants of our ancestor's beliefs cling tightly to the modern mind. Some, like the fear of heights, arise from the adapted feeling that 'this is not a safe place to be,' and can benefit us enormously. Others, like paranoia, arising from a soft whispering or the rustling of a bush, can force us into unnecessarily anxiety. Joseph Conrad's short story *An Outpost of Progress* draws interesting conclusions about similar feelings of inclusion and exclusion:

> Few men realise that their life, the very essence of their character, their capabilities and their audacities, are only the expression of their belief in the safety of their surroundings. The courage, the composure, the confidence; the emotions and principles; every great and every insignificant thought belongs not to the individual but to the crowd: to the crowd that believes blindly in the irresistible force of its institutions and of its morals, in the power of its police and of its opinion. But the contact with pure unmitigated savagery, with prime nature and primitive man, brings sudden and profound trouble into the heart. To the sentiment of being alone of one's kind, to the clear perception of the loneliness of one's thoughts, to one's sensations – of the negation of habitual, which is safe, there is added the affirmation of the unusual, which is dangerous; a suggestion of things vague, uncontrollable and repulsive, whose discomposing intrusion excites the imagination and tries the civilised nerves of the foolish and the wise alike.[67]

Strengthening D'Arcy Thompson and Alan Turing's tradition of mathematical biology (now generally known as biomathematics), there is also evidence which suggests that Watson and Crick were wrong in claiming that the emergence of the genetic code is purely an accident, and that its appearance has 'a hidden rationale that is an expression of natural law. Structures deriving from the principles of symmetry braking are apparently at the heart of much of our understanding of pattern formation in the living world, and also in language.'[68] As biolinguistics has emphasised, patterns of structural perfection exist both in brain anatomy and in grammar.[69] Discrete infinity (yielded through FLN, possibly recursion and Merge) may also 'have far deeper roots than functional utility in organizational principles at work in other combinatorial systems in both biology and physics, rather than being specifically adapted for language and communication,' systems of 'unbounded diversity of form and function [depending] on a combinatorial hierarchy arising through the combining and permutation of discrete elements drawn from a finite set,' including genes, atoms, ions, molecules or natural numbers.[70] Observing similar associations, Hinzen notes that the

> rational numbers necessarily "contain" or "entail" the natural, and the reals the rationals, in such a way that the mappings of the formal objects involved to the right part-whole relations do not have to be stipulated, but follow from the dimentionality of the formal arithmetical system itself. The right "entailments" in the clausal domain – that a VP contains an NP, and a vP a VP – might follow in a similar fashion.[71]

On the dimensionality of the formal arithmetic system, Elizabeth Stapel elaborates that, after defining the naturals, wholes, integers and rationals, 'each new type of number contained the previous type within it. The wholes are just the naturals with zero thrown in. The integers are just the wholes with the negatives thrown in. And the fractions are just the integers with all their divisions thrown in.'[72] Insofar as the reals entail the rational and the rationals entail the naturals, it is quite an open playing-field, though it has some limits, since 'in the eight-dimensional algebraic spaces inhabited by special numbers called octonions, standard algebraic operations such as associativity cease to be defined.'[73]

Certain constraints imposed by the conceptual-intentional system will also impact the structure of language. The nature of concepts and their categorization specifically influences lexical semantics, as Chater and Christiansen point out. Citing other influences, they add that 'the infinite range of possible thoughts presumably is likely to promote tendencies toward compositionality in natural language; the mental representation of time is likely to have influenced linguistic systems of tense and aspect; and more broadly, the properties of conceptual structure may profoundly and richly influence linguistic structure.'[74] Many human-specific concepts are also hierarchic 'lexical groupings that allow ever-increasing levels of abstraction (e.g., Cat → Animal → Living thing → Object).'[75]

The attention of biolinguistics on Lewontin's 'Triple Helix' (of environment, genes and the organism), Gould's 'historical,' 'functional' and 'formal' causal influences on the creation of natural objects, and Chomsky's 'three factors' of data (exposure to language-specific data), genetic endowment and natural law leads to a familiar approach in biology; namely 'focus on the internal system and domain-specific constraints; data analysis of external events has its place within these constraints.'[76] We should not ignore the possibility (even likelihood) that there exists 'fourth factor' effects currently unknown and which may altogether lie beyond the grasp of physical inquiry. The goal of the scientist, then, is not to rush to adaptationist stories to account for the structure of an organic system, but to 'see how much of the complexity of an organism you can explain in terms of general properties of the world. ... Insofar as there is a residue, you have to attribute it to some specific genetic encoding; and then you've got to worry about where that came from.'[77] This biolinguistic approach is in many ways an updated version of Turing's biomathematical view that 'we must envisage a living organism as a special kind of system to which the general laws of physics and chemistry apply ... and because of the prevalence of homologies, we may well suppose, as D'Arcy Thompson has done, that certain physical processes are of very general occurrence.'[78] Wallace Arthur termed these 'third factor' constraints 'bias' when writing on the directed development and evolution of embryos.[79] If language is as organic as embryos and not 'some non-spatial, non-temporal phantasm' (Wittgenstein), any methodological naturalist should focus on similar 'bias' when studying the growth and evolution of FLN.[80] One such 'third factor' is principles of computational efficiency, like why dividing cells during mitosis form spheres and not cubes; others include principles of efficient computation not specific to language, such as principles of locality and structure preservation.

These principles of minimal search, Chomsky believes, are responsible for the labelling phenomenon in language which assigns a construction its 'head' (a key word determining the properties of the construction, with the phrase 'excited about October' being an adjectival phrase in which the head is 'excited'), with Merge being a head and another object. In this view, 'a computational system that embodies minimal search will isolate one element of a combined pair (i.e. roughly a binary-branching phrase) as more accessible than the other. This element is the head.'[81] Endocentricity ('headedness,' a feature arithmetic lacks) for Chomsky is consequently a product of minimal search and an essential aspect of a 'perfect' computational system. Cedric Boeckx differs from this view in proposing that endocentricity is a language-unique property:

> [Labelling], as far as I can tell, is very, very specific to language as a kind of hierarchical structure. If you look elsewhere in other systems of cognition (music, planning, kinship relations, etc.) you find a lot of evidence for hierarchical structuring of systems, possibly recursive ones, but as far as I can tell, those hierarchical structures are not headed or endocentric in the same way that linguistic structures are. That, to my mind, is very specific to language, so while you find hierarchies everywhere, headed or endocentric hierarchies seem very central to language.[82]

Contrary to connectionist and usage-based theories, 'the core generative procedures [of language] seem to have a character not explained by domain-general learning.'[83] In the study of poverty of the stimulus, Berwick, Pietroski, Yankama and Chomsky recommend that anyone serious about the evolution and nature of language should follow the lead of the minimalist program, since

> by taking POS [poverty of the stimulus] arguments seriously, theorists are led to a question that turns out to be fruitful: To what degree would the simplest imaginable procedures for generating unboundedly many meaningful expressions be procedures that give rise to the observed phenomena of constrained ambiguity? ... In our view, the way forward begins with the recognition that environmental stimuli radically underdetermine developmental outcomes, and that grammar acquisition is a case in point. The one can try to describe the gap between experience and linguistic knowledge attained, reduce that gap to basic principles that reflect the least language-specific innate endowment that does justice to the attained knowledge, and thereby help characterize the true role of experience in a manner that illuminates cognition.[84]

This innate knowledge 'come[s] from the organism in the form of "core" domains, skeletal principles, biases, or conceptual assumptions.'[85] Rochel Gelman also captured the spirit of the generative enterprise by stating in her introduction to the 1990 Special Issue of *Cognitive Science* that

> Experience is indeterminant or inadequate for the inductions that children draw from it in that, even under quite optimistic assumption about the nature and extent of the experiences relevant to a given induction, the experience is not, in and of itself, sufficient to justify, let alone compel, the induction universally drawn from it in the course of development. For example, there is nothing in the environment that supports a child's conclusion that the integers never end.[86]

Linguistics may also not concern itself with its present neurobiological validity, as Berwick and co stress, and choose to focus on its own theories – a common occurrence in the history of the sciences. Newton, for instance, 'had made the theory of some such universal attractive force necessary, by laying completely aside his unripe and vague conjectures as to the material cause of attraction, and kept strictly to what he could prove – the mathematical causes of the phenomena, supposing that there were some principle of approximation operation inversely as the square of the distance, let its physical nature be what it may.'[87] In the case of Newton's 'occult force,' 'the mathematical construction went ahead of the physical explanation, and on this occasion the circumstance was to attain a significance unsuspected by Newton himself.'[88] A primary 'occult force' in biolinguistics, the basic combinatorial operation of language, has been outlined by Chomsky in his 'Three Factors' paper:

> An elementary fact about the language faculty is that it is a system of discrete infinity. Any such system is based on a primitive operation that takes *n* objects already constructed, and constructs from them a new object: in the simplest case, the set of these *n* objects. Call that operation Merge. Either Merge or some equivalent is a minimal requirement. With Merge available, we instantly have an unbounded system of hierarchically structured expressions. The simplest account of the 'Great Leap Forward' in the evolution of humans would be that the brain was rewired, perhaps by some slight mutation, to provide the operation Merge, at once laying a core part of the basis for what is found at that dramatic 'moment' of human evolution, at least in principle; to connect the dots is no trivial problem. There are speculations about the evolution of language that postulate a far more complex process: first some mutation that permits two-unit expressions (yielding selectional advantage in overcoming memory restrictions on lexical explosion), then mutations permitting larger expressions, and

finally the Great Leap that yields Merge. Perhaps the earlier steps really took place, but a more parsimonious speculation is that they did not, and that the Great Leap was effectively instantaneous, in a single individual, who was instantly endowed with intellectual capacities far superior to those of others, transmitted to offspring and coming to predominate, perhaps linked as a secondary process to the SM system for externalization and interaction, including communication as a special case. At best a reasonable guess, as are all speculations about such matters, but about the simplest one imaginable, and not inconsistent with anything known or plausibly surmised. In fact, it is hard to see what account of human evolution would not assume at least this much, in one or another form.[89]

This small neural re-wiring, occurring in an individual, would have created a computational procedure, the minimum required being Merge, which takes two lexical items and constructs an extended object: x and y merge as the set $\{x, y\}$:

Merge: Target two syntactic objects α and β, form a new object Γ $\{\alpha,\beta\}$, the label LB of $\Gamma(LB(\Gamma)) = LB(\alpha)$ or $LB(\beta)$.[90]

Merge was then linked to pre-existing conceptual structures (with primate cognition being far from primitive). This would initially have been a Language of Thought (LOT), later being linked to the sensory-motor system for externalisation, since the ability to plan and externalise thoughts clearly would have conferred adaptive benefits. What these conceptual structures are exactly remains a mystery, with the nature of even the simplest lexical items being incredibly rich – perhaps too rich for a physical theory of semantics to ever emerge (a point to be developed below). To take a case from contemporary biolinguistics, the nature of anaphoric 'binding' may be determined by innate constraints which predate the emergence of language, and many aspects of it may be semantic in nature: The sentence 'John painted a portrait of himself,' for instance, is based on the semantic representation of 'portrait' ('a representation of John') and not syntactic factors.[91] And contrary to Pinker and Jackendoff's stress on adaptation:

UG's properties cannot be the work of NS [natural selection] for there is reason to think that the human faculty of language is about 50–100,000 years old at most. Conclusion: The basic architecture of the Faculty of Language (FL) is not the result of an arduous and long process of selection but reflects an at most small addition to an already existing system of computations.[92]

The mutation yielding the neural re-wiring and Merge would have occurred in an individual, not a group. Since there would have been no selectional pressures acting on the mutation, it follows that it would have developed in a form dictated by natural law (like a snowflake, to use Chomsky's example) – a 'perfect' system satisfying the demands imposed by the pre-existing conceptual-intentional interface, and resulting in a Language of Thought with no externalization.[93] Later on the individual would have faced the non-trivial task of relating two unrelated cognitive systems, the C/I-interface and the perceptual-articulatory (P/A)-interface, the latter having likely existed for hundreds of thousands of years. The Norman Invasion and the Great Vowel Shift, then, led to drastic changes in externalization (with 'French' and 'Latin' loan words pouring into the macaronic Middle English language), but not in the syntax or semantics. Explaining such a transfer of loan words 'is probably a job for the ordinary perspective' of language (E-language) – obeying a strictly cognitive (or neural) perspective will tell us as much about this phenomenon as the principles of quantum mechanics will tell us anything about the Battle of Waterloo.[94] With the relation between the narrow faculty of language and this pre-existing conceptual structure remaining in obscurity, Hinzen is amongst the biolinguists who assume that 'the human language faculty provides *forms* that a possible human structural meaning may have, leaving a residue of non-structural meanings (concepts), a substantive amount of which we share with other animals that lack syntax (or at least do not use it, or do not use it for the purposes of language).'[95] Unbounded Merge may not have added concepts, then, but rather allowed us to structure them, which otherwise exist in non-humans but are 'encapsulated in particular cognitive domains.'[96] On the importance that laws of nature may play on the determination of the structure and growth of an organism, Hinzen points out that 'there are certain operations that the structure of our nervous system, perhaps by virtue of deeper principles of a mathematical and physical nature that is instantiates, makes available; and that they lend themselves to certain uses, as in discourse or quantification.'[97]

Of the two central kinds of Merge available, 'Internal Merge' is responsible for transformations and displacement (movement) while 'External Merge' is responsible for phrase-structure building. But as to why syntactic objects get transformed to begin with, Hinzen points out that it may be a meaningless question:

> A transformation is a "warping" of a phrase marker (a phrase structure generated by the base component), in a topological sense, and just as there is no "reason" for a topological operation

in an n-dimensional mathematical space, there is maybe no basis for looking for one in our case. Things in nature, or in the spaces of mathematics, do not move or get transformed for a reason: as Galileo argued, answering why-questions is not part of the scientific endeavour.[98]

Merge could also have yielded the emergence of mathematics, which possibly 'piggybacked' off language. If Merge is constrained to a unary operation it is constrained to the successor function, which yields the theory of natural numbers. The successor function would have, along with many other things, left us susceptible to the Uncaused Cause argument in theology; that every created thing must have a thing which created it (3 cannot equal 4 without having 1 added to it). Chomsky's minimalism is also similar to von Neumann's theory of ordinals, which builds them recursively from a single object, such as the empty set $\emptyset = \{\}$. As Hinzen points out, 'Applying Merge qua set-formation to this single object, it yields the singleton set $\{\emptyset\}$, a set different from the empty set in containing exactly one object, the empty set itself. Since Merge is recursive, it can be applied again to that second object, $\{\emptyset\}$, to yield a set that contains that very set $\{\emptyset\}$ plus its single member: $\{\emptyset, \{\emptyset\}\}$.'[99] We can repeat this procedure ad infinitum or instead use Neumann's infinite ordinal ω, much like the natural numbers (a recursive process similar to Church's lambda-calculus and Peano-Dedekind arithmetical induction). Arithmetic, McGilvray points out, is thus 'a product of internal Merge operating over a lexicon with a single element, and it is a very impoverished natural language.'[100] Merge, in this view, is domain-general. Noting further similarities between syntax, number theory and topology, Kuroda recently discovered that 'there exists a formal procedure for transforming the Euler product representations of certain ζ-functions (a fundamental concept in number theory) into phrase-structure representations, an intriguing result that should be readily translated into the Merge-based generative system.'[101] The neurological diversity between the localization of language and mathematics also doesn't rule out the possibility that arithmetic is yielded from unbounded Merge. As Luigi Rizzi has pointed out, any dissociation in the capacities following brain lesions relate purely to the use (performance) or these capacities, and not the possession of them (competence).[102]

Hinzen also identifies the semantic component of language to be 'the most obscure, at least if we look at it from a naturalistic or biolinguistic perspective.'[103] 'Theories of semantics,' he adds, 'are about a domain in nature that we do not understand.'[104] From early generative grammar, from the 1970s up to the Government and Binding framework, and on through to the Minimalist Program of contemporary biolinguistics, Hinzen notices that the models of generative grammar's architecture have seen 'a consistent *downward shift* of "semantic representation", corresponding mostly to the philosophical notion of a "thought". We began with (the semantic representation of) "Thought" being at the top, but we finish with it being at the bottom. First, it predates syntax (is an output to it), then it is its product, or something that is at least partially *explained by the derivational process* itself.'[105] In Chomsky's 1965 *Aspects of the Theory of the Syntax*, semantic interpretation was read off 'Deep Structure,' governed by phrase structure rules operating on lexical items (LEX):

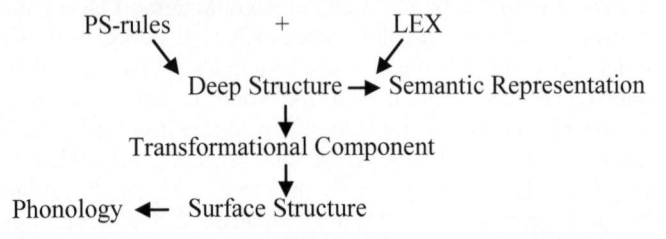

```
PS-rules         +          LEX
         ↘              ↙
         Deep Structure → Semantic Representation
                ↓
         Transformational Component
                ↓
Phonology ← Surface Structure
```

By 1995, the architecture of generative grammar was confined to the lexicon, the structure-building rule Merge/Move, with the S-structure of Government and Binding being replaced with the operation 'Spell-Out,' which 'strips off the phonetic features of a representation from its semantically interpretable ones,' yielding Phonetic Form (PF) and Logical Form (LF):[106]

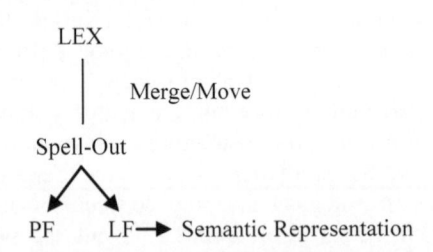

```
            LEX
             |    Merge/Move
         Spell-Out
           ↙  ↘
         PF    LF → Semantic Representation
```

Hinzen brings the development of the generative models to their 'logical conclusion: semantic interpretation is now done as an inherent part of the dynamically proceeding derivation itself.' He details how

> as a semantic representation is beginning to be constructed, Spell-Out applies several times, as an inherent part of a derivation that itself proceeds in stages or "cycles", and the interfaces are accessed at each of these points. An LF-representation is never assembled as such; neither are the other old levels of representations, D-structure and S-structure.

Instead of such *grammar-internal* levels of representation, we now merely have the semantic interface with the Conceptual-Intentional System, SEM, which, as an interface, is not grammar-internal, and is also accessed several times in the course of the derivation.

His final model of minimalist architecture is detailed as follows:

```
    LEX
      \
       \——— Spell-Out    PF
        \
         \——— Spell-Out    PF
          \
           \      etc.
            \
            (SEM)
```

In this radical minimalist theory

> there is no semantic component, no independent generative system of "thought," no "mapping" from the syntax to such a system, no semantic "interface." There is a computational system (syntax), which constructs derivations; periodically, after each "phase" of a computation, the generated structure is sent off to the sensorimotor systems; and there are no structured semantic representations beyond the ones that the syntax is inherently tuned to construct.[107]

Meaning is clearly something computed in a derivation, as this lowering of semantic representations during the history of generative grammar shows. But meaning in the truth-conditional, logical sense may consequently not be a suitable object for naturalistic inquiry, since the internal structure of 'Logical Form' (LF) and the SEM (C/I) interface are biological objects, like RNA, as opposed to artefacts. Drawing on similar architecture, Berwick, Pietroski, Yankama and Chomsky have recently outlined the biolinguistic view of language's evolution in these terms:

> Modern logic, which has been intertwined with the study of arithmetic notions, has revealed various ways of describing generative procedures. But one way or another, such procedures require a primitive combinatory operation that forms larger elements out of smaller ones. (The primitive operation can be described in terms of a Peano-style axiom system, a Fregean ancestral, a Lambek-style calculus with "valences," or in other

ways.) We assume that some aspect of human biology implements a primitive combinatory operation that can form complex expressions from simpler ones, which themselves may or may not be atomic. Call this basic operation – whatever its source, and without further stipulations about how it should be characterized – merge.

But despite all this, as Berwick has pointed out, it is sometimes argued that the above

> recursive generative capacity is somehow *necessary* for communication, thus bridging the gap between protolanguage and recursive human syntax. But is this so? There seem to be existing human languages that evidently possess the ability to form recursive sentences but that apparently do not need to make use of such power: a well-known example is the Australian language Warlpiri, where it has been proposed that a sentence that would be recursively structured in many other languages, such as "I think that John is a fool" is formed via linear concatenation, 'I ponder it. John is a fool'; or to take a cited example, *Yi-rna wita yirripura jaru jukurrpa-warnu wiinyiinypa*, literally, "little tell-PRESENT TENSE story dreaming hawk," translated as "I want to tell a little dreaming story about a hawk". Evidently, then, recursion is not essential to express "the beliefs about the intentional states of others," quite contrary to what some researchers such as Pinker and Bloom and more recently Pinker and Jackendoff have asserted.[108]

In an essay entitled 'All you Need is Merge,' Berwick also speculates that 'Chimpanzees, and possible other primates, got conceptual atoms. Birds got rhythm [from FOXP2]. But only people combined both, getting undeletable features, *and* Merge. And from this, came language: Merge is all you need.'[109] Conversely, Jackendoff and Pinker stress a view of the mind which holds it to be 'intrinsically functionally designed. The human linguistic mind, in particular, is thought [by them] to be the result of the external shaping of selective forces, acting on the communicative functions and other effects of subcomponents of the language system as a whole, eventually composing it, piece by piece, in a gradualistic fashion.'[110] But Jackendoff and Pinker's essays do not pose a threat to biolinguistics. If anything, they strengthen its claims, since

> it is a coherent possibility at this moment of inquiry that there is *no* mechanism within human language use that is unique to language and unique to humans, *including* the crucial element of recursion. If so, all that is new in humans is how these

mechanisms get *integrated* into a system serving some novel function, in which case the need for adaptationist explanation of the mechanisms involved in that system falls apart.[111]

As with much of evolutionary biology, Frances Egan sides with Jackendoff and Pinker and 'presume[s] that computational mechanisms are *adaptations*; organisms have such mechanisms because they enhanced fitness in the ancestral environment. But being adapted to an environment is itself a contingent feature of a computational mechanism, so regarded. It is coherent to imagine the same (type of) computational mechanism being built by IBM or coalescing fully formed out of a swamp.'[112] According to a more recent adaptationist proposal, 'the fit between the neural mechanisms supporting language and the structure of language itself is better explained in terms of how language has adapted to the human brain, rather than vice versa. This solution to the logical problem of language evolution, however, requires abandoning the notion domain-specific UG.'[113]

Dennett and Pinker's choice of 'God or natural selection' also misses the point entirely. God may even be a more sensible option considering the adaptationist's explanation of cognition, since the functionalist doctrine articulated by Pinker (*'how well something works* plays a causal role in *how it came to be'*[114])

> flies in the face of the Darwinian doctrine that neither adaptation nor natural selection are sources of genetic change. Mutations are the sources of novelty, their causes are internal, and they are crucially undirected, hence not intrinsically functional. Functionality helps to explain novel organic forms in the sense of explaining their selective retention or elimination, and the changes in their distribution within a population over evolutionary time. Explaining how new forms arise in the first place on the level of the individual organisms is a totally different matter. As it stands, Pinker's phrase simply conflates function with genesis.[115]

These Spencerian distortions (which have also helped shift attention away from the effective founder of sociobiology and anarchist Peter Kropotkin's response to Spencer, his theory of 'mutual aid,' lying in direct opposition both to social Darwinism and state-corporate culture) have been revived by Dennett, who regards Spencer as 'an important clarifier of some of Darwin's best ideas,' despite the fact that Darwin himself favoured a Kropotkin-style view of human nature.[116] Though Dennett rightly criticises the actual applications of social Darwinism, his distortions have had much influence in the mystical and peculiarly irrational sub-discipline of the philosophy of mind. Again ignoring the rich history of formalist and biomathematical thought, functionalists like Pinker, Dennett and Dawkins regard natural selection as either the major or only force in evolution, leading Hinzen to conclude that Natural Theology is not much different from their idea of evolution.

Even in evolutionary psychology, where Pinker's ideas have been welcomed most, 'the very structure of particular mental organs is *posited* on the basis of an assumption about adaptive history. The dependence of form on function is a matter of methodological principle, and it seems unclear why we can permit ourselves this methodology when studying the mind, though we apparently cannot when studying physiology (another methodological dualism).'[117] Pinker and Dennett are thus foundationalists in seeking origins and causes and not the causes of form, making 'no attempt to actually trace the effects of natural selection of language design, hence to derive the structural patterns we find from whatever "function" language is stipulated to have.'[118]

With the Neo-Darwinian adaptationists exhibiting clear shortcomings, a more insightful set of viewpoints from which to examine the organic world was constructed by G. C. Williams, who considered three perspectives: the organism-as-document, focusing on evolutionary history and viewing organisms as the result of discrete contingencies; the organism-as-artefact, which considers organisms as having evolved complex designs for specific purposes; and organism-as-crystal, which 'views the design of the organism as the outcome of a structure-building process which involves laws of form and natural constraints that induce restrictions in the space of logically possible designs and force nature to generate only a tiny amount of a much larger spectrum of forms.' Analogous to the language faculty,

> The design of a snowflake exhibits an inherent geometry, a sixfold symmetry. In explaining it, extraneous factors like temperature, humidity, or air pressure will play a role. However, to say that these external factors *explain* the internal structure would be as strange as saying that the water we give to a plant causes it to develop in the way it does. Just as the plant's development will be a matter of what species of plant it is, and of internally directed principles of growth for which external factors are merely necessary conditions, the crystal is primarily explained internalistically, by laws of form.[119]

Similar arguments, incidentally, were given by Chomsky in his debate with Foucault against the notion that there was no fixed human moral nature; namely, that even if Foucault were right, there would still remain the empirical question of what cognitive faculties entered into the interpretation of experience and its production of moral judgments (even Locke at times endorsed the a priori nature of moral instincts), with these faculties themselves constituting what has come to be known as our moral grammar (or MG).[120] A sensible conceptual foundation for inquiry into our moral faculties was laid by Hume (and subsequently ignored by Parisian postmodern 'philosophers'): He noticed that since we're capable of making an infinite number of moral judgments in a boundless number of scenarios, it follows that there must be some finite (and innate, in his view, handed to us by 'the original hand of nature') moral principles, with MG generating certain judgments about experience. The leading expert on this 'internalist' perspective of morality, John Mikhail, whose PhD thesis revived Rawls' ideas, explains that his program to construct a Universal Moral Grammar uses models similar to Chomskyan linguistics in an attempt to answer traditional philosophical questions about morality.[121] Touching upon both domains, he points out how 'every natural language seems to have words or phrases to express basic deontic concepts, such as *obligatory*, *permissible*, and *forbidden*, or their equivalents,' and that 'individuals are intuitive lawyers who are capable of drawing intelligent distinctions between superficially similar cases, although their basis for doing so is often obscure.'[122] Yet 'differences between the moral and the linguistic faculty nonetheless abound, with the former for example being subject to learning, instruction, and moral conflict in a way that linguistic judgements are not.'[123] It should be seen as quite uncontroversial (in fact necessary) that this set of innate principles forms part of human nature. Rejecting a different form of relativism from Foucault's, Chomsky counters Rorty's position

> which holds in an extreme form that apart from their basic physical structure humans have no nature. They have only history, and their thought can be modified without limit. Nothing like this can be even close to true if taken literally. ... No one doubts that a person's understanding, judgments, values and goals reflect acquired cultures, norms, conventions and so on. But these are not mind-external entities. They are not acquired by taking a pill. They are constructed by the mind on the basis of scattered and constructed experiments. And they are constantly applied in circumstances that are novel and complex.[124]

But as Hinzen points out, there is 'a kind of consensus in the humanities today that the very idea of human nature is a historical relic from the eighteenth-century Enlightenment, of which the nineteenth century quite rightfully deprived us, and which the twentieth century rightly took care not to re-establish, was it not for the sociobiologist's unfortunate revival of it.'[125] Even much of contemporary biology is averse to the notion of 'human nature,' seeing itself as 'a deeply historical science.'[126] Hinzen cites David Hull as falling prey to a purely organism-as-document perspective:

> If species are interpreted as historical entities, then particular organisms belong in a particular species because they are part of that genealogical nexus, not because they possess any essential traits. No species has an essence in this sense. Hence there is no such thing as human nature. There may be characteristics which all and only extant human beings possess, but this state of affairs is contingent, depending on the current evolutionary state of *Homo sapiens*. Just as not all crows are black (not even potentially), it may well be the case that not all people are rational (even potentially).[127]

There is very little to support 'the sweeping Neo-Darwinian picture of human nature that Singer, Dennett, or Dawkins defend and make the basis of their vision on ethics. It is a completely empirical question whether the apparent absence of functional design and external shaping we have found in language and concepts carries over to human morals.'[128] There is no reason to study morality from a purely functionalist perspective, assuming with Pinker that moral concepts are 'designed by natural selection to further the long-term interests of individuals and ultimately their genes.'[129] If morality developed predominantly through internal constraints and not environmental pressures then 'human nature may become a *positive* constraint on ethical thinking, not merely the negative one it is in [Singer's *A Darwinian Left*]. That is, it could be a *source* of human morality, not merely a deplorable *limitation* to it.'[130] As for the rise of the New Atheist crew (supporting such self-appointed guardians of global virtue as the late neoconservative and chief propagandist for the 2003 invasion of Iraq, Christopher Hitchens – 'a hero of the mind' as Stephen Fry recently called him, with his usual melodrama), Chomsky comments:

> Many of these people [sociobiologists and Neo-Darwinist], like Dawkins, regard themselves very plausibly as fighting a battle for scientific rationality against creationists and fanatics and so on. And yes, that's an important social activity to be engaged in, but not by misleading people about the nature of evolution – that's not a contribution to scientific rationality. Tell them the truth about evolution, which is that selection plays some kind of a role, but you don't know how much until you know.[131]

It's also important to stress our vast ignorance of how language evolved in the first place. McGilchrist, reviewing the neuroscientific literature of the past few decades, concludes: 'Music is likely to be the ancestor of language and it arose largely in the right hemisphere, where one would expect a means of communication with others, promoting social cohesion, to arise.' Although this speculation lacks any evidence and virtually any conceptual justification (it's most likely the other way around), he goes on make the perceptive point that 'we have lost the sense of the central position that music once occupied in communal life, and still does in most parts of the world today.' The origins of organised music and dance will have significant bearing on our later explorations of religion, since all of these topics (including language itself) relate in intricate ways. All cultures have a rich variety of music, but we in the West

> have relegated music to the sidelines of life. We might think of music as an individualistic, even solitary experience, but that is rare in the history of the world. In more traditionally structured societies, performance of music plays both an integral, and an integrative, role not only in celebration, religious festivals, and other rituals, but also in daily work and recreation; and it is above all a shared performance, not just something we listen to passively. It has a vital way of binding people together, helping them to be aware of shared humanity, shared feelings and experiences, and actively drawing them together. In our world, competition and specialisation have made music something compartmentalised, somewhere away from life's core.[132]

With this in mind, McGilchrist's view of music is that it 'is, of all the arts, the one that is most dependent on the right hemisphere; of all aspects of music, only rhythm is appreciated as by the left hemisphere, and it may not be accidental that, while contemporary art music has become the preserve of a few devotees (in a way that was never previously true of new music in its time), popular music in our age has become dominated by, and almost reduced to, rhythm and little else.'[133] Neurologist Oliver Sacks, writing in 2006, expressed similar anxieties in noting that the 'primal role of music if to some extent lost today, when we have a special class of composers and performers, and the rest of us are often reduced to passive listening. One has to go to a concert, or a church or a music festival, to recapture the collective excitement and bonding of music. In such a situation, there seems to be an actual binding of nervous systems.'[134]

McGilchrist also rightly dispels certain misconceptions about the left hemisphere being used for reason and language with the right hemisphere specialising in creativity and emotion – they use both sides of the brain: 'every single brain function is carried out by both hemispheres. Reason and emotion and imagination depend on the coming together of what both hemispheres contribute.'[135] Nevertheless he stresses, as many others have, that language is where the left hemisphere 'is at home,' since it 'has a much more extensive vocabulary than the right, and more subtle and complex syntax. It extends vastly our power to map the world and to explore the complexities of the causal relationships between things.'[136]

Though McGilchrist himself doesn't see it this way, I think (along with the biolinguistic stress on the sudden emergence of the 'narrow faculty of language,' Merge) this makes some sense out of the fact that poetry came before prose: 'Prose was at first known as *pezos logos*, literally "pedestrian, or walking, *logos*", as opposed to the usual dancing *logos* of poetry. In fact early poetry was sung: so the evolution of literary skill progresses, if that is the correct word, from right-hemisphere music (words that are sung), to right-hemisphere language (the metaphorical language of poetry), the left-hemisphere language (the referential language of prose).'[137] Supporting Chomsky's belief that language is primarily for 'beauty' and the expression of thought, Peter Ludlow notes that the selectional advantage of language 'might be something as simple as having a system that can generate certain metrical patterns – in effect for creating poetry.'[138] For Plato and Aristotle, 'the sense of wonder is the mark of the philosopher – philosophy indeed has no other origin,' and 'it is owing to their wonder that men both now begin, and at first began, to philosophize.'[139]

Whether music or language had great utility or not, their existence makes at least one thing obvious; namely that we 'can *choose* to imitate forms of thinking or behaving; and by so doing both speed up our evolution by many order of magnitude, and shift it away from the blind forces of chance and necessity, in a direction of our own choosing.'[140] The very fact that we have developed 'useless' and 'futile' human universals like dancing and smiling 'tells us a great deal about our security within a relatively secure, well-off group, who had the time to maintain useless habits.'[141] But we cannot choose to get rid of syntax or recursion, or still more mysterious notions like those capacities which enter into decision-making and willing. They are, as Kleist understood, species-defining properties:

> "Grace appears purest in that human form which has either no consciousness or an infinite one, that is, in a puppet or in a god."
>
> "Therefore," I said, somewhat bewildered, "we would have to eat again from the Tree of Knowledge in order to return to the state of innocence?"

"Quite right," he answered. "And that's the last chapter in the history of the world."[142]

REFERENCES

[1] Lucretius, *One the Nature of the Universe*, trans. Ronald Melville (Oxford University Press, 2008), Book I, l. 146-9, p. 7.
[2] Wolfram Hinzen, *Mind Design and Minimal Syntax* (Oxford University Press, 2006), p. 71.
[3] Julian Huxley, *Birds and the Territorial System* (London: Chatto & Windus, 1926).
[4] Hinzen, *Mind Design and Minimal Syntax*, p.74.
[5] Paul Dirac, 'Methods in theoretical physics,' *From a Life of Physics; Evening lectures at the International Centre for Theoretical Physics, Trieste, Italy*, A. Salam et al., A special supplement of the International Atomic Energy Agency Bulletin, Austria.
[6] Robert Freidin and Jean-Roger Vergnaud, 'Exquisite connections: some remarks on the evolution of linguistic theory,' *Lingua*, 111, 2001: 639-66.
[7] Robert Sklar, 'Chomsky's revolution in linguistics,' *The Nation*, 9 September 1968.
[8] Noam Chomsky, *The Science of Language: Interviews with James McGilvray* (Cambridge University Press, 2012), p. 23.
[9] Noam Chomsky, 'Some notes on economy of derivation and representation,' *MIT Working Papers in Linguistics*, 10, 1989: 43-74.
[10] Hiroki Narita and Koji Fujita, 'A Naturalist Reconstruction of Minimalist and Evolutionary Biolinguistics,' *Biolinguistics*, 4(4), 2010: 385 (356-76). See Juan Uriagereka, 'Multiple spell-out,' *Working Minimalism*, eds. Samuel David Epstein and Norbert Hornstein (MIT Press, 1999), pp. 251–282.
[11] Noam Chomsky, *Aspects of the Theory of Syntax* (MIT Press, 1965), p. 6; see Eric Lenneberg, *Biological Foundations of Language* (New York: John Wiley & Sons, 1967).
[12] Andrew Radford, *Syntactic Theory and the Structure of English: A Minimalist Approach* (Cambridge University Press, 1997), p. 6.
[13] Richard Feynman, *The Feynman Lectures in Physics*, vol. 1 (Reading, Mass.: Addison-Wesley, 1963), p. 26.
[14] Cited in Cedric Boeckx, *Linguistic Minimalism: Origins, Concepts, Methods, and Aims* (Oxford University Press, 2006), p. 113.
[15] Isaac Newton, *Philosophiae Naturalis Principia Mathematica* (London, 1687), ii. 160ff.
[16] Francis Bacon**Error! Bookmark not defined.**, *The Works*, trans. and ed. J. Spedding, R. L. Ellis and D. D. Heath, 7 vols., (London: Longmans, 1858 [1857-9]), Book I, aphorism X, p. 48.
[17] Cited in Boeckx, *Linguistic Minimalism*, p. 117.
[18] Nicolaus Copernicus, *On the Revolutions of the Heavenly Spheres*, R. M. Hutchins (ed.), *Great Books of the Western World* (Chicago: Encyclopaedia Britannica, 1952), Book I.8, p. 519.
[19] Cited in Chris Harman, *A People's History of the World: From the Stone Age to the New Millennium* (London: Verso, 2008), p. 145.
[20] Jean-Jacques Rousseau, *Emile: Or, On Education* (1762).
[21] Bertrand Russell, *The Problems of Philosophy* (Oxford University Press, 1959/1912), p. 93.

[22] Lawrence Sklar, 'Convention, Role of,' *A Companion to the Philosophy of Science*, ed. W. H. Newton-Smith (Oxford: Blackwell, 2001), p. 58.
[23] Frank Jackson, *Perception: A Representation Theory* (Cambridge University Press, 1977, repr. Gregg Revivals, 1993), p. 120.
[24] Hinzen, *Mind Design and Minimal Syntax*, p. 28.
[25] Daniel Dennett, *Darwin's Dangerous Idea* (New York: Simon & Schuster, 1995), p. 225.
[26] Michael Gazzaniga, *Conversations in the Cognitive Sciences* (MIT Press, 1997), p. 114.
[27] Evelyn Fox-Keller, *Making Sense of Life* (Harvard University Press, 2002), p. 74, p. 87.
[28] Anna Maria Di Sciullo and Cedric Boeckx, 'Introduction: Contours of the Biolinguistic Research Agenda,' *The Biolinguistic Enterprise: New Perspectives on the Evolution and Nature of the Human Language Faculty*, ed. Anna Maria di Sciullo and Cedric Boeckx (Oxford University Press, 2011), p. 2; cited in Harold Bloom, *The Anxiety of Influence: A Theory of Poetry*, 2nd ed. (Oxford University Press, 1997), p. 42; see Steven Pinker and Paul Bloom, 'Natural Language and Natural Selection,' *The Behavioral and Brain Sciences*, vol. 13 (1990): 704-84.
[29] Marx X. Wartofsky, *Conceptual Foundations of Scientific Thought: An Introduction to the Philosophy of Science* (New York: Macmillan, 1968), p. 29.
[30] Richard Dawkins, 'The Origins of the Specious,' *The Independent*, 27 July 1994.
[31] Richard Dawkins, *The Oxford Book of Modern Science Writing* (Oxford University Press, 2009), p. 30.
[32] Bruce W. Weber and David J. Depew, 'Developmental systems, Darwinian evolution, and the unity of science,' *Cycles of Contingencies: Developmental Systems and Evolution*, eds. Susan Oyama, Paul E. Griffiths and Russell D. Gray (MIT Press, 2001), pp. 239–253.
[33] Stephen Jay Gould, *The Structure of Evolutionary Theory* (Harvard University Press, 2002), p. 290.
[34] Aaron T. Goetz, Todd K Shackelford and Steven M. Platek, 'Introduction to evolutionary psychology: A Darwinian approach to human behavior and cognition,' *Foundations in Evolutionary Cognitive Neuroscience*, eds. Steven M. Platek and Todd K. Shackelford (Oxford University Press, 2009), p. 6 (1-21).
[35] J. Tooby and L. Cosmides, 'The psychological foundations of culture,' *The Adapted Mind: Evolutionary Psychology and the Generation of Culture*, eds. J. H. Barkow, L. Cosmides and J. Tooby (Oxford University Press, 1992), p. 62 (19-136).
[36] August Weismann, 'The All-Sufficiency of Natural Selection,' *Contemporary Review*, 64: 309-38, 596-610.
[37] William James, 'Great Men and Their Environment,' Lecture before the Harvard Natural History Society, *Atlantic Monthly*, October 1880; Michael C. Corballis, 'The Evolution of Consciousness,' *The Cambridge Handbook of Consciousness*, p. 588.
[38] Boeckx, *Linguistics Minimalism*, pp. 9-10.
[39] Ibid., p. 10.
[40] Stuart A. Newman, Gabor Forgacs and Gerd D. Müller, 'Before programs: The physical origination of multicellular forms,' *International Journal of Developmental Biology*, 50, 2006: 290 (289-99).
[41] Hinzen, *Mind Design and Minimal Syntax*, p. x.
[42] Edward Stuart Russell, *Form and Function* (London: John Murray, 1916).
[43] See R. Amundson, 'Typology reconsidered: Two doctrines on the history of evolutionary biology,' *Biology and Philosophy*, 13, 1998: 153-77.
[44] Charles Darwin, *The Descent of Man* (Sioux Falls, South Dakota: NuVision Publications, LLC, 2007), p. 84.

[45] George Gaylord Simpson, *The Meaning of Evolution* (Yale University Press, 1949), pp. 291-292.
[46] Daniel Sennert, *Tractatus de consensus et dissensu Galenicorum et Peripateticorum cum Chymicis, Opera omnia*, vol. 3 (Lyons, 1650), p. 765.
[47] Wolfram Hinzen, *An Essay on Names and Truth* (Oxford University Press, 2007), p. 14.
[48] L. T. Hobhouse, *Mind in Evolution* (London: Macmillan, 1901), pp. 11-12.
[49] Hinzen, *Mind Design and Minimal Syntax*, p. 15.
[50] Brian K. Hall, *Evolutionary Developmental Biology*, 2nd ed. (Norwell, MA.: Kluwer Academic Publishers, 1999), p. 74.
[51] George McGhee, *Theoretical Morphology* (Columbia University Press, 1998), p. 2.
[52] Ray Jackendoff, *Language, Consciousness, Culture: Essays on Mental Structure* (Massachusetts: MIT Press, 2007), p. 73.
[53] Richard Levins and Richard Lewontin, *The Dialectical Biologist* (Harvard University Press, 1985), p. 89.
[54] M. J. West and A. P. King, 'Settling nature and nurture into an ontogenetic niche,' *Developmental Psychobiology*, 20, 1987: 552 (549–562).
[55] Chomsky, *The Science of Language*, p. 246.
[56] Cedric Boeckx, 'Round Table: Language Universals: Yesterday, Today, and Tomorrow,' *Of Minds and Language: A Dialogue with Noam Chomsky in the Basque Country*, ed. Massimo Piatello-Palmarini, Juan Uriagereka and Pello Salaburu (Oxford University Press, 2009), pp. 195-6 (195-220). See Kurt Lewin, *A Dynamic Theory of Personality* (New York: McGraw-Hill, 1935).
[57] Galileo Galilei, *Letter to Madame Christina of Lorraine, Grand Duchess of Tuscany: Concerning the Use of Biblical Quotations in Matters of Science* (1615), trans. Stillman Drake, www.disf.org/en/documentation/03-Galileo_Cristina.asp.
[58] Leonardo Olschki, 'Galileo and the Scientific Revolution,' *Philosophical Review*, 211, 1943: 363.
[59] Cited in Hinzen, *Mind Design and Minimal Syntax*, p. 105.
[60] Ernst Mayr, *What Makes Biology Unique?* (Cambridge University Press, 2004), p. 69, 28, 93, cited in Boeckx, *Linguistic Minimalism*, p. 133.
[61] E. Pennis, 'Evo-Devo enthusiasts get down to details,' *Science*, 298(5595), November 2002: 953-5.
[62] Christopher Cherniak, 'Innateness and brain-wiring optimization: Non-genomic nativism,' *Cognition, Evolution, and Rationality*, ed. A. Zilhao (London: Routledge, 2005).
[63] T. H. Huxley, *Man's Place in Nature and Other Essays* (London: Macmillan, 1901), p. 151.
[64] Theodosius Dobzhansky, *Mankind Evolving* (Yale University Press, 1962).
[65] Steven Pinker, *The Stuff of Thought: Language as a Window into Human Nature* (London: Penguin, 2008), p. 104.
[66] Russell, *The Problems of Philosophy*, pp. 11-12.
[67] Joseph Conrad, *Heart of Darkness and Other Tales*, ed. Cedric Watts (Oxford University Press, 2008), p. 5.
[68] Hinzen, *Mind Design and Minimal Syntax*, p. 94. See A. Moro, *Dynamic Antisymmetry* (MIT Press, 2000).
[69] Christopher Cherniak, 'Innateness and brain-wiring optimization: Non-genomic nativism,' *Cognition, Evolution, and Rationality*, ed. A. Zilhao (London: Routledge, 2005).
[70] Hinzen, *Mind Design and Minimal Syntax*, p. 130.
[71] Ibid., p. 193.
[72] Elizabeth Stapel, 'Number Types,' Purplemath, http://www.purplemath.com/modules/numtypes.htm.

[73] Hinzen, 'Hierarchy, Merge, and Truth,' p. 132, n. 5.
[74] Chater and Christiansen, 'Language Acquisition Meets Language Evolution,' 1136. See T. Suddendorf and M. C. Corballis, 'The evolution of foresight: What is mental time travel, and is it unique to humans?', *Behavioral and Brain Sciences*, 30:299-351.
[75] Vladimir M. Sloutsky, 'From Perceptual Categories to Concepts: What Develops?', *Cognitive Science*, 34(7), September 2011: 1245 (1244-86).
[76] Stephen Jay Gould, *The Structure of Evolutionary Theory* (Harvard University Press, 2002); Berwick et al., 'Poverty of the Stimulus Revisited,' 1209.
[77] Chomsky, *The Science of Language*, p. 132.
[78] Alan Turing, 'The Chemical Basis of Morphogenesis,' *Philosophical Transactions of the Royal Society of London B*, 237(641), August 1952: 37-72.
[79] Wallace Arthur, *Biased Embryos and Evolution* (Cambridge University Press, 2004).
[80] Ludwig Wittgenstein, *Philosophical Investigations*, § 108.
[81] Alex Drummond and Norbert Hornstein, 'Basquing in Minimalism,' review of *Of Minds and Language*, ed. Massimo Piatelli-Palmarini, Juan Uriagereka and Pello Salaburu, *Biolinguistics*, 5(4), 2011: 353 (347-65).
[82] Boeckx, 'The Nature of Merge: Consequences for Language, Mind, and Biology,' *Of Minds and Language*, pp. 47-8.
[83] Berwick et al., 'Poverty of the Stimulus Revisited,' 1225.
[84] Ibid., 1239.
[85] Vladimir M. Sloutsky, 'Mechanisms of Cognitive Development: Domain-General Learning or Domain-Specific Constraints?', *Cognitive Science*, 34(7), September 2010: 1126 (1125-30).
[86] Rochel Gelman, 'Structural Constraints on Cognitive Development: Introduction to a Special Issue of Cognitive Science,' *Cognitive Science*, 14(1), January 1990: 4 (3-10).
[87] Friedrich Albert Lange, *The History of Materialism and Criticism of its Present Importance*, trans. Ernest Chester Thomas, 3rd ed. (London: Routledge & Kegan Paul, 1957), p. 308.
[88] Ibid., p. 309.
[89] Noam Chomsky, 'Three Factors in Language Design,' *Linguistic Inquiry*, 36(1), Winter 2005: 11-12 (1-22).
[90] Noam Chomsky, *The Minimalist Program* (MIT Press, 1995).
[91] See Ray Jackendoff and Peter Culicover, *Simpler Syntax* (Oxford University Press, 2009).
[92] Drummond and Hornstein, 'Basquing in Minimalism,' 351.
[93] See Noam Chomsky, 'Governing Board Symposium with Noam Chomsky: The Biology of Language in the 21st Century,' 33rd Annual Cognitive Science Conference, July 20-23 2011, Boston, Massachusetts, http://thesciencenetwork.org/programs/cogsci-2011/governing-board-symposium-with-noam-chomsky.
[94] Ray Jackendoff, *A User's Guide to Thought and Meaning* (Oxford University Press, 2012), p. 17.
[95] Hinzen, *Mind Design and Minimal Syntax*, p. 235.
[96] Wolfram Hinzen, *An Essay on Names and Truth* (Oxford University Press, 2007), p. 52.
[97] Hinzen, *Mind Design and Minimal Syntax*, p. 214.
[98] Ibid., p. 208.
[99] Hinzen, *Mind Design and Minimal Syntax*, p. 190.
[100] Chomsky, *The Science of Language*, p. 182.
[101] Hiroki Narita and Koji Fujita, 'A Naturalist Reconstruction of Minimalist and Evolutionary Biolinguistics,' *Biolinguistics*, 4(4), 2010: 358 (356-76). See S. Y. Kuroda, 'Suugaku to seiseibunpoo: "Setumeiteki datoosei-no kanatani" sosite gengo-no suugakuteki jituzairon [Mathematics and generative gram mar: "Beyond explanatory

adequacy" and mathematical realism of language] (with an extended English summary), *Sophia Linguistica*, 56, 2009: 1–36.
[102] Luigi Rizzi, *Some Elements of the Study of Language As a Cognitive Capacity* (London: Routledge, 2003).
[103] Wolfram Hinzen, 'Emergence of a Systematic Semantics through Minimal and Underspecified Codes,' *The Biolinguistic Enterprise*, p. 417.
[104] Ibid., p. 419.
[105] Hinzen, *Mind Design and Minimal Syntax*, p. 156.
[106] Ibid., p. 157.
[107] Wolfram Hinzen, 'Hierarchy, Merge, and Truth,' *Of Minds and Language*, p. 128 (123-141).
[108] Robert C. Berwick, 'Syntax Facit Saltum Redux: Biolinguistics and the Leap to Syntax,' *The Biolinguistic Enterprise*, p. 71. See Steven Pinker and Paul Bloom, 'Natural Language and Natural Selection,' *The Behavioral and Brain Sciences*, vol. 13 (1990): 704-84; Steven Pinker and Ray Jackendoff, 'The Faculty of Language: What's Special About It?', *Cognition*, vol. 95 (2005): 201-36.
[109] Robert C. Berwick, 'All you Need is Merge: Biology, Computation, and Language from the Bottom Up,' *The Biolinguistic Enterprise*, p. 491.
[110] Hinzen, *Mind Design and Minimal Syntax*, pp. 12.
[111] Ibid., p. 170.
[112] Frances Egan, 'Naturalistic Inquiry: Where does Mental Representation Fit In?', *Chomsky and His Critics*, p. 97.
[113] Nick Chater and Morten H. Christiansen, 'Language Acquisition Meets Language Evolution,' *Cognitive Science*, 34(7): 1133 (1131-57).
[114] Steven Pinker, *How the Mind Works* (New York: W. W. Norton & Company, 1999), p. 162.
[115] Hinzen, *Mind Design and Minimal Syntax*, p. 96-7.
[116] Daniel Dennett, *Darwin's Dangerous Idea: Evolution and the Meaning of Life* (New York: Simon & Schuster, 1995), p. 393.
[117] Hinzen, *Mind Design and Minimal Syntax*, p. 103.
[118] Ibid., p. 107.
[119] Ibid., p. 12. See G. C. Williams, *Natural Selection: Domains, Levels, and Challenges* (Oxford University Press, 1992).
[120] John Locke, *An Essay Concerning Human Understanding*, ed. A. D. Woozley (Glasgow: William Collins Sons & Co Ltd, 1964, repr. 1984), Book IV, ch. 4, sect. 7.
[121] John Mikhail, 'Rawls' Linguistic Analogy: A Study of the 'Generative Grammar' Model of Moral Theory Described by John Rawls in 'A Theory of Justice,' Phd Dissertation (Cornell University, 2000).
[122] John Mikhail, 'Universal moral grammar: theory, evidence, and the future,' *Trends in Cognitive Sciences*, 11(4), April 2007: 143, 150 (143-52).
[123] Wolfram Hinzen, Nirmalangshu Mukherji and Bijoy Boruah, 'The Character of Mind,' *Biolinguistics*, 5(3), 2011: 280 (274-83).
[124] Noam Chomsky, 'Language and the Mind Revisited: Language and the Rest of the World,' The Charles M. and Martha Hitchcock Lectures, *UC Berkeley Graduate Council Lectures*, July 2003, http://www.youtube.com/watch?v=_tvPkSveevA.
[125] Hinzen, *Mind Design and Minimal Syntax*, p. 32.
[126] Stuart Kauffman, *At Home in the Universe: The Search for Laws of Complexity* (London: Penguin, 1995), p. 22.
[127] David Hull, 'A Matter of Individuality,' *Philosophy of Science*, 45, 1978: 358 (335-60).
[128] Hinzen, *Mind Design and Minimal Syntax*, p. 276.
[129] Steven Pinker, *How the Mind Works* (New York: W. W. Norton, 1997), p. 406.
[130] Hinzen, *Mind Design and Minimal Syntax*, p. 277.

[131] Chomsky, *The Science of Language*, p. 105.
[132] McGilchrist, *The Master and His Emissary*, p. 104.
[133] Ibid., p. 418.
[134] Oliver Sacks, 'The power of music,' *Brain*, 2006, 129(10): 2528-32.
[135] Interview with Natasha Mitchell, 19 June 2010, 'The Master and his Emissary: the divided brain and the reshaping of Western civilisation,'
http://www.abc.net.au/rn/allinthemind/stories/2010/2928822.htm.
[136] McGilchrist, *The Master and His Emissary*, p. 70.
[137] Ibid., p. 105.
[138] Peter Ludlow, *The Philosophy of Generative Linguistics* (Oxford University Press, 2011), p. 36.
[139] Plato, *Theaetetus*, trans. J. McDonald (Oxford: Clarendon, 1973), 155d 2ff; Aristotle, *Metaphysics*, trans. W. D. Ross, 2 vols. (Oxford: Clarendon, 1924), A 2, 982b 11ff.
[140] McGilchrist, *The Master and His Emissary*, p. 124 (emphasis his).
[141] R. I. M. Dunbar, *The Human Story: A New History in Mankind's Evolution* (London: Faber, 2004), p. 131.
[142] Heinrich von Kleist, 'On the Marionette Theatre,' trans. C.-A. Gollub, in A. L. Wilson (ed.), *German Romantic Criticism* (New York: Continuum, 1982), p. 244.

3

SCOPES AND LIMITS

I was about to put the final period to these notes, just as the ancients put crosses over the pits where they had thrown their dead, when suddenly the pencil shook and dropped from my fingers.
    'Listen.' I tugged at my neighbour. 'Just listen to me! You must – you must give me an answer: out there, where your finite universe ends! What is out there, beyond it?'
    He had no time to answer. From above, down the stairs – the clatter of feet...
YEVGENY ZAMYATIN[1]

Space: what you damn well have to see.
JAMES JOYCE[2]

It's a common assumption both among physicists and philosophers to assume that the modern scientific revolution, beginning in the seventeenth century, endowed humans with an unbounded cognitive reach into nature's mysteries. The contrary view, derided in philosophical circles as 'mysterianism,' holds that humans do in fact have cognitive 'scopes and limits' (Russell) which impose constraints on what we can discover about the internal workings of the mind and of natural laws. But it is not a law of nature (as far as we know), nor is there anything in the theory of evolution which suggests that humans should be able to answer the questions we ask of the world. Nor is it clear that we are asking the right questions to begin with. Empirical naturalistic inquiry is certainly the most successful 'method' we have, but it's important to stress that scientific understanding is 'a kind of chance convergence between aspects of the natural world and properties of the human mind/brain, which has allowed some rays of light to penetrate the general obscurity.'[3] Or rather than a 'nature versus nurture' debate, what is needed is a 'nature via nurture' debate.[4] Colin McGinn concurs that 'Perception is essentially an ongoing link between the world and our knowledge of it.'[5] The perceptual world (or 'phenomenal world,' as Koffka put it in his classic *Principles of Gestalt Psychology*) is the reality constructed by our 'cognoscitive powers' in response to the external world.[6] Biological organisms simply aren't in the business of directly representing the external world molecule-for-molecule. Constrained by the shaping effects of physical laws on cognitive systems (as Alan Turing and D'Arcy Thompson emphasised), they instead construct a model of the world reliable enough to survive, plan action and propagate genes. To the distress of the physicists, evolution doesn't seem too concerned with how the world *really* is. Tricks such as seeing a 'square' when present with four dots arranged in a particular position work well for survival and (along with language) creativity, and 'It is only in the context of the laboratory that their artificiality is detected.'[7]

This is also true for cheetahs and ladybirds, and unless humans are angels (to use Chomsky's wry phrase), we should expect ourselves to have cognitive scopes and limits as well (putting a nail firmly in the coffin of Ban van Fraassen's belief that 'there are no science stoppers'[8]), though this is something which many philosophers deny, like Daniel Dennett and his partner-in-Reason and casual imperialist Sam Harris (he of the poetic *Letter to a Christian Nation* and similar apologetics for criminal and destructive Western 'intervention' in the Middle East). And Diderot's wonderful frankness reminds us that, for all our abstractions, we are nevertheless part of the biological world: 'Underlying our most sublime sentiments and our purest tenderness there is a little of the testicle.'[9] Locke was one of the first to focus on these epistemological and cognitive concerns, trying to divert questions of the mind away from metaphysics:

> When we know our own strength, we shall the better know what to undertake with hopes of success; and when we have well surveyed the *powers* of our own minds, and made some estimate what we may expect from them, we shall not be inclined either to sit still, and not set our thoughts on work at all, in despair of knowing anything, nor, on the other side, question everything, and disclaim all knowledge, because some things are not to be understood.[10]

In words Dobzhansky would have appreciated, Darwin wrote in one of his early notebooks: 'Origin of man now proved. Metaphysics must flourish. He who understands baboon would do more towards metaphysics than Locke.'[11] To give an example of this truism, recent research has suggested that the development of bipedalism was a breakthrough in the origins of laughter. Kate Douglas writes that 'four-legged mammals must synchronise their breath with their stride. By taking pressure off the thorax, bipedalism gave us the breath control needed for speaking and the ability to chop up our exhalations, giving the characteristic ha-ha-ha sound of human laughter.'[12]

The limitations of our mathematical faculties could easily disallow us from solving, among others, the infamous P = NP problem. But mathematics is a much simpler case than our conceptual structures: 'Our minds have managed an analysis of the generative principle of number – an easy case – but they fail badly in the analysis of *house*, which is already much too complex, apparently, and maybe there is not much hope that we will ever succeed for something like *justice*.' Putting this disparity down to 'principled cognitive limitations,' Hinzen notes that 'There are reasons for pessimism in this domain ... which there may not be in the domain of the mathematical sciences, whose central concepts and modes of understanding are easier.'[13] Bennett and Hacker's *Philosophical Foundations of Neuroscience* suggests with Hinzen that 'Investigating logical relations among concepts is a philosophical task. Guiding that investigation down pathways that will illuminate brain research is a neuroscientific one.'[14]

Michael Lockwood, whose words should act as a counter to the Dennett-lead denial of phenomenology, adds that consciousness 'provides us with a kind of "window" on to our brains, making possible a transparent grasp of a tiny corner of material reality that is in general opaque to us. ... The qualities of which we are immediately aware, in consciousness, precisely *are* at least some of the intrinsic qualities of the states and processes that go to make up the material world – more specifically, states and processes within our own brains. This was Russell's suggestion.'[15] Again, contra Dennett, there is in fact cognitive phenomenology (comprehending mathematics, metaphor and so on) as well as sensory phenomenology. The encompassing term 'consciousness,' mostly a wreckage site today, has also been 'forced through the terminological looking glass by philosophers like Dennett who use it to mean precisely something that involves no consciousness.'[16] We are still 'virtually mute' about the nature of experiential content, notes Chomsky, with Galen Strawson adding in *Mental Reality* that 'experiential phenomena outrun the resources of human language in a way that makes a human science of experiential phenomena look impossible.'[17]

In the same year David Chalmers' baptised the problem of consciousness 'The Hard Problem,' Piet Hut and Roger Shepard asked 'But what is a thing? When we look carefully, then we find that what we considered to be an object appears in our consciousness as a bundle of meanings, draped around sense impressions that are far, far less complete and filled in and filled up than the 'real thing' we feel to be present, three-dimensionally, continuous in time. ... Its reality? Nothing but a sense of reality.'[18] Considering this self-generating 'sense of reality' a primary aspect of consciousness, David Geary speculates that the 'critical differences comparing humans to other species appears to be in the ability of [generating] representations of the self in working memory – self awareness – and to mentally time travel.' Considering this, he concludes an essay on fluid intelligence with the following suggestion:

> Although climatic and ecological conditions can create variation and novelty and thus may have contributed to the evolution of mental time travel and the ability to form self-aware mental simulations of potential future states, the most dynamic and variable conditions faced by most humans are those that arise from the competing interests of other people. My proposal is that the evolved function of these mental models is to generate a self-centred simulation of the "perfect" world, one in which other people behave in ways consistent with one's best interest, and biological and physical resources are under one's control. The function of mental simulations is to create and rehearse strategies that can be used to reduce the difference between this perfect world and current conditions. The cognitive systems that evolved to support the use of these self-centred mental models are known as working attentional control, that is, the core cognitive components of general fluid intelligence.[19]

There are doubtless many other factors involved in the generation of self-awareness, like the urge to inquire and create and Hardy's 'instinct for co-operation,' but Greary's general proposals could well be on the right track (putting aside their Neo-Darwinian groundings). Similarly, François Jacob speculated that the mind needs to have 'a representation of the world that is unified and coherent,' with folk science and religion offering an unshakeable grip on it, since 'mythical explanation often does better than scientific explanation. For science does not aim at reaching a complete and definitive explanation of the whole universe,' instead idealising to domains, seeking 'partial and provisional answers for certain phenomena that can be isolated and well defined.'[20]

The Russian mathematician Andrey Kolmogorov seems have been thinking along the same lines: 'The human brain is incapable of creating anything which is really complex.' William G. Lycan adds that 'there is nothing the mind does that calculators and automobile engines do not do, albeit on a smaller scale.'[21] The brain sciences have mapped the computational aspects of language to certain areas of the brain, but the creative aspect eludes explanation, and we still don't fully understand how events in the world register in the mind: 'The poetic character of thinking,' as Heidegger said, 'is still veiled over.'[22] In his Russell

Lectures at Cambridge University in 1971, Chomsky made the following crucial point about biological restrictions and creativity:

> The image of a mind, initially unconstrained, striking out freely in arbitrary directions, suggests at first glance a richer and more hopeful view of human freedom and creativity, but I think that this conclusion is mistaken. Russell was correct in titling his study *Human Knowledge: Its Scope and Limits*. The principles of mind provide the scope as well as the limits of human creativity. Without such principles, scientific understanding and creative acts would not be possible. If all hypotheses are initially on a par, then no scientific understanding can possibly be achieved, since there will be no way to select among the vast array of theories compatible with our limited evidence and, by hypothesis, equally accessible to the mind. One who abandons all forms, all conditions and constraints, and merely acts in some random and entirely willful manner is surely not engaged in artistic creation, whatever else he may be doing. "The spirit of poetry, like all living powers, must of necessity circumscribe itself by rules," Coleridge wrote, perhaps "under laws of its own origination." If, as Russell frequently expressed it, man's "true life" consists "in art and thought and love, in the creation and contemplation of beauty and in the scientific understanding of the world," if this is "the true glory of man," then it is the intrinsic principles of mind that should be the object of our awe and, if possible, our inquiry.[23]

Chomsky also makes the Russellian distinction between 'mysteries' and 'problems.' Mysteries are beyond our cognitive reach, and problems (like prime number puzzles) are not. A mystery for a bee not be a mystery for us, and conversely. Or, as Michael Tanner puts it in *Schopenhauer*, 'our experiences are the result of a collaboration between us and a basic reality of which we can know nothing, except that it must exist.'[24] Rationality, for Wartofsky, is 'suited for one fairly short and atypical set of cosmic circumstances.'[25] There's certainly nothing in the human genome which suggests that we should be able to answer the questions we ask. This is not particularly profound, but an important fact which is too often ignored. Contrary to the claims of the postmodernists (if their 'intellectual rubbish,' to borrow Russell's phrase, even rises to the level of 'claims'), Hume took it for granted that any view of morality, politics and aesthetics is founded (consciously or not) on certain assumptions about human nature: '*Mathematics*, *Natural Philosophy*, and *natural Religion*, are in some measure dependent on the science of Man; since they lie under the cognizance of men, and are judged of by their powers and faculties.'[26] Strawson adds an important qualification: The claim that the world has a fixed nature

> is not threatened by the suggestion that our best models of the behaviour of things like photons credit them with properties that seem incompatible to us – wave-like properties and particle-like properties, for example. What we learn from this is just that this is how photons affect us, given their intrinsic nature – given how they are in themselves, and we are in ourselves. We acquire no reason to think (incoherently) that photons do not have some intrinsic nature at any given time. Whatever claim anyone makes about the nature of reality – including the claim that is had apparently incompatible properties – just is a claim about the way it is. This applies as much to Everett's "many worlds" theory of reality as to any other.[27]

Strawson's words also ring true in the mind of Heisenberg: 'we have to remember that what we observe is not nature in itself but nature exposed to our method of questioning.'[28] These concerns date back to Socrates, who thought it absurd to study the heavens before we understand ourselves, since a clear definition of the tools used to examine the heavens can yield insights into the nature of the heavens themselves. Niels Bohr also suggested that 'It is wrong to think that the task of physics is to find out how nature *is*. Physics concerns what we can say about nature.'[29] The task of cognitive science, we might add, concerns the faculties 'termed mental' (Priestley) which determine what it is possible to say at all. Pinker takes the view that 'the minds we are stuck with harmonize with the world well enough for science to be possible.'[30] This 'convergence' between the mind and the world was also appreciated by Hume who, taking in his Newtonian surroundings, comments in the *Enquiry* that 'when we talk of gravity, we *mean* certain effects, without ever comprehending *that active power*.'[31] '[T]he cause of Gravity ....,' he later confessed, 'I do not pretend to know.'[32] Hume also laid the foundations for a cognitive explanation of Descartes' contact mechanics, rejecting the need for common sense to accord with any '*active power*': '[W]hen we say ... that one object is connected with another, we *mean only* that they have acquired a connexion in our thought ... conclusion which is somewhat extraordinary, but which seems founded on sufficient reason.'[33] When Strawson's photons behave in incomprehensible ways to us, we have only to read Friedrich Lange to set us straight:

> [The] speculating individual ... lives in the world of his mental cobwebs, and contrives to make everything harmonise with them. Before a thing can have really become a thing to *him*, it

must have modelled itself upon his ideas. But all things are not so yielding, and experience plays such philosophers the awkwardest tricks.[34]

The search for nature's laws may be tiresome, he adds:

Yet our efforts will never be wholly in vain. The truth, though late, yet comes soon enough; for mankind will not die just yet. Fortunate natures hit the right moment; but never has the thoughtful observer the right to be silent because he knows that for the present there are but few who will listen to him.[35]

Out intuitive contact mechanics, the source of Locke's 'motion,' is founded on the concept of force dynamics, which turns out to be

very different from our best understanding of force and momentum from Newtonian physics. The force-dynamic model in language singles out one entity and conceived of another as impinging on it, whereas in physics neither object in an interaction is privileged. Language conceives of the agonist [the object in motion under the influence of an internal or external force] as having an inner impulse toward motion or rest, whereas physics treats an object as simple continuing at its current velocity. Language distinguishes motion and rest as qualitatively distinct tendencies, whereas physics treats rest as a velocity that happens to be zero. ... In language, things can just happen without stated causes – *The book toppled off the shelf*; *The sidewalk cracked* – whereas in physics every event has a lawful antecedent.[36]

To pick an example almost at random, *New Scientist* recently reported that 'it doesn't look like quantum mechanics will ever be intuitive. Take away one of its weirdest components – the uncertainty principle – and you end up with a perpetual motion machine.'[37] Aristotle's folk physics of motion requiring contact was also demolished by Newton's 'occult' laws: To the question 'Why does steam rise and a rock fall?' Aristotle answered in his *Metaphysics* by claiming that they're seeking their natural place. It was only when later 'natural philosophers' like Galileo allowed themselves to be puzzled by things which seemed obvious did modern science begin (as Chomsky stresses). Embracing our own puzzlement about apparently obvious things – like why rocks fall and why we can add one to an indefinite series of natural numbers – is often the beginning of science. Sticking with folk science certainly won't help with naturalistic investigations. The way we use the word 'energy' ordinarily is different from the way physicists use it – the various connotations of the term 'energy' are ignored in the physics department (the same goes for 'water' in the chemistry department, differing radically from H2O 'physically' and conceptually, contrary to contemporary dogma in the philosophy of language – a point to be returned to below):

> As scientists come to understand the target phenomenon [the external world] in greater depth and detail, they highlight the aspects of the metaphor that ought to be taken seriously and pare away the aspects that should be ignored. ... The metaphor evolves into a technical term for an abstract concept that subsumes both the target phenomenon and the source phenomenon. It's an instance of something that every philosopher of science know about scientific language and that most laypeople misunderstand: scientists don't "carefully define their terms" before beginning an investigation. Instead they use words loosely to point to a phenomenon in the world, and the meanings of the worlds gradually become more precise and the scientists come to understand the phenomenon more thoroughly.[38]

Notice that anything we come to understand sufficiently, with some degree of confidence we call 'physical' (while leaving the ordinary notion of 'physical,' such as 'tangible' and 'perceptual,' aside). Anything we don't understand is simply part of 'the world.' The words of the novelist and anarchist Edward Abbey describe the attitude of pre-Newtonian 'natural philosophers' and of many thinkers today: 'Whatever we cannot easily understand we call God; this saves much wear and tear on the brain tissues.'[39] But since our understanding of 'physical' keeps changing (with new particles being discovered and explanatory theories being revised), mystics like Deepak Chopra (the Jacques Derrida of popular science) are speaking nonsense in describing their profound insights into nature as the workings of 'non-physical' forces (perhaps even residing in divine entities lying 'outside' of space and 'beyond' time – i.e. the exact definition of nonexistence). Lucretius' answer to these 'immaterial' forces may have sounded something like the following:

> And if you cannot explain the things you see
> Without inventing tiny parts of matter
> Endowed with the same nature as the whole,
> This reasoning puts an end to all your atoms.
> They'll simply shake their sides and rock with laughter,
> And salt tears run in rivers down their cheeks.[40]

All of this has repercussions for the notion of free will, which is commonly understood to be the ability to do 'whatever they will or please, both with respect to the operations of their minds and the motions of their bodies, uncontrolled by any foreign principle or cause.'[41] 'Elective decision' for Schopenhauer, is deeply embedded in our experience of the world. As is widely known, an experiment by Benjamin Libet demonstrated that 'if you ask people to make voluntary movements, their brains initiate the movement before they become consciously aware of any intention to move.'[42] The reason, then, why free will feels so free is because it is the result of a mixture of voluntary and involuntary actions. This mismatch between unconscious decision and conscious awareness leads Jeffrey Gray to speculate that 'perhaps there is something about the processes that create consciousness that do in fact require it to operate slowly.'[43] 'It may be,' said Turing in 1951, shortly before his tragic death, 'that the feeling of free will which we have is an illusion. Or it may be that we really have got free will, but yet there is no way of telling from our behaviour that this is so' – suggesting, like Stuart Hameroff, that our biology simply gives us no choice but to believe in free will, 'real' or not.[44]

But physics, on the other hand, is commonly claimed to do the opposite. Take the two main areas of study and publication in the physical and mathematical sciences over the last few decades, both with alleged repercussions for free will and consciousness: determinacy and randomness. There's been much of work done on both of these, but maybe there are aspects of the world which don't fall within either of them (human will, action and language are used appropriately to situations but are not caused by them, as Chomsky often notes: a genuine paradox in the sciences, yet to be resolved). If we can't explain such things, perhaps we are simply revealing our cognitive scope and limits. With an apparent jump of their imaginations, the determinists (like Dennett and Harris) may object soberly that such things as free will 'contradict what physics tells us about the world': an interesting but false claim, presupposing an understanding of what 'physical' is, as mentioned ('But learning just as certainly that his will is subject to laws, he does not and cannot believe it'[45]). For Hume (and for modern science) we are 'ignorant of the manner or force by which a mind, even the supreme mind, operates either on itself or on the body.'[46] The study of the visual system also concerns solely what happens between the retina and the striate cortex, exploring only mechanisms. Neurons in area V3 of the primate brain, for instance, respond to moving shapes, those in V4 respond to colours, and those in the middle temporal area (MT) and V5 are specialised to respond to motion.[47] But nobody even bothers to ask the question 'Why did Walt look at Jesse?' Rather, the brain and cognitive sciences leave the hard questions of decision and will aside and focus on the mechanisms which enter into decision-making and action. The visual system happens to be highly modular (or 'composite'), with the ventral and dorsal pathways processing objects differently (justifying the claims of nineteenth-century anatomists that the visual system is a heterogeneous construction). The visual system must also interact with conceptual and memory systems, leading O'Regan and Noë to suggest that a sensory modality is a specialised 'mode of exploration' of the world 'mediated by distinctive sensori-motor contingencies.'[48] But there are no scientific explanations as to why Walt looked at Jesse. We may use intentional language to explain it in human terms ('He looked at him because he was angry'), but these descriptions ignore the fact that '"Mental causation" (whatever that is) is quite different from mechanical causation; we don't know how they relate.'[49] As Jeffrey Gray points out, 'the physiology of the neuromuscular junction, by which the brain causes muscle fibres to contract, is one of the best understood components of nervous activity,' although, again, nobody even attempts to ask why an individual contracted their biceps at one point and not another.[50]

For a different example, take the nematode caenorhabditis elegans (C. elegans), a roundworm which has only 300 neurons (compared with our 80-100 billion). Neuroscientists like Sydney Brenner and Craig Mello (Nobel laureates) know exactly how it functions, but nobody knows why it turns left rather than right. It's simply too complex an organism. By the time humans are reached on the scale of neural complexity, questions of will and decision are up in the stratosphere, far away from the horizon of rational inquiry. The computational neuroscientist Sebastian Seung, in his recent study *Connectome*, points out that 'if we stick to the manual methods used for the roundworm, it would take a million person-years to reconstruct the wiring in a cubic millimetre of the human brain's cortex.'[51]

The classical liberals, from Adam Smith to Jacques Rousseau to Wilhelm von Humboldt, believed that the crucial, distinguishing essence of man was his freedom: 'The defining features of the human condition can all be traced to our ability to stand back from the world, from our selves and from the immediacy of experience. This enables us to plan, to think flexibly and inventively, and, in brief, to take control of the world around us rather than simply respond to it passively. This distance, this ability to rise above the world in which we live, has been made possible by the evolution of the frontal lobes.'[52] It's in a similar spirit to this that John Dupré holds that 'There are no known physical laws that predict when a cat is going to go off hunting or a dog decide to chase its tail, and it is no more than dogma that such laws must exist.'[53] Anthony Gottlieb echoes Jackendoff in stressing that 'there is a growing realisation among some neuroscientists that looking at flickers of activity inside our heads can be a misleading way to see how our minds work. This is because many of the distinctively human things that people do take place over time and outside their craniums. Perhaps the brain is the wrong place to look if you want to find free will.'[54] Even Benjamin Libet's experiments only show that some decisions are made unconsciously; hardly a refutation of free will (recall Nietzsche's 'small terse fact that a thought comes when "it" wishes and not when "I" wish').[55] Judging reaction times is only one experimental method, and a limited one at that. Challenging the current vogue of denying the existence of this species-defining capacity, Chomsky has recently pointed out that

> there is no reason to believe that even if decisions of an organism are unconscious, that they are determined. To have reason to believe this, one would have to have in hand a decent naturalistic theory of an organism's decision-making, and getting that is extremely unlikely. ... [T]he mind's various systems operate relatively independently and have multiple inputs and outputs. Constructing a deterministic theory of an organism's action/behavior would require coming up with a solution to an n-system problem. This is a far more complex task than coming up with a solution for the extremely simple cases found with the systems discussed in n-body problems. There are far more variable factors, assuming even that we had a completed list of the mental systems involved and their specific contributions.[56]

Rejecting the Searlean proposal that mental 'rules' have to be 'accessible to consciousness' (another widespread dogma in philosophy) in favour of a Nietzschean view of thought, Strawson also writes:

> In large-scale bodily action, Davidson remarks, "we never do more than move our bodies; the rest is up to nature". In cognition we never do more than aim or tilt our minds; the rest is up to nature, trained or not. Much bodily movement is ballistic [in the psychophysical sense in which the movement of a leg is ballistic after the required neural activity has occurred], relative to the initiating impulse; the same goes for thought.[57]

The following processes, writes Jeffrey Gray, 'are capable of being, and normally are, completed unconsciously before we have any conscious awareness of what is being carried out: analysis of sensory input; analysis of emotional content of input; phonological and semantic analysis of heard speech; semantic and phonological preparation of one's own spoken words and sentences; learning' the formation of memories; choice and preparation of voluntary acts; planning and execution of movements.'[58] The scarcely understood adaptive functions of self-awareness have also been outlined by Thomas Metzinger:

> There is an internal image of yourself that you cannot recognize *as* an image while it is there – and the unromantic part is in regarding this as a weapon that emerged in the cause of the cognitive arms race. There was constant competition among organisms on this planet, for millions of years, and it was merciless and cruel and the development of things like memory, better perceptions, was just as important as better legs, better lives, better hearts. I like to look at the human self-model as a neurocomputational weapon, a certain data structure that the brain can activate from time to time, such as when you have to wake up in the morning and integrate your sensory perceptions with your motor behaviour. The ego machine just turns on its phenomenal self, and that is the moment when *you* come to.[59]

Russell, who knew the sciences better than most philosophers, also speculated with Darwinian plausibility that

> All perennial controversies, such as that between determinists and believers in free will, spring from a conflict between opposing passions, both widespread, but one stronger in one man and the other in another. In this case, the conflict in between the passion for power and the passion for safety, because if the external world behaves according to law we can adapt ourselves to it. We desire the reign of law for the sake of safety, and freedom for the sake of power.[60]

This is not to argue that there are no other ways to understand the world. It's hard to deny that more can be understood about human nature by reading novels or studying history than the natural sciences will likely ever reveal. This may be due to our cognitive structure, but outside of very narrow domains naturalistic inquiry rarely proves itself capable of explanation. In the same way, Shelley's verse and the rest of 'the arts may offer appreciation of the heavens to which astrophysics does not aspire':[61]

> Art thou pale for weariness
> Of climbing heaven and gazing on the earth,
> Wandering companionless

> Among the stars that have a different birth, –
> And ever changing, like a joyless eye
> That finds no object worth its constancy?[62]

Colin McGinn writes how the drama of Shakespeare illustrates 'the human mind as we recognize it – as we experience it in the marketplace and at home.'[63] There is a great difference between experiencing the marketplace and knowing precisely what areas of the brain are responsible for that experience. To illustrate, the theoretical physicist Lee Smolin, in his careful and perceptive study *The Life of the Cosmos*, discusses the levels of abstraction the mathematician is forced to deal with in order to glimpse the constants of natural law. He is of the opinion that mathematics

> is responsible for the obscurity that surrounds the creative processes of theoretical physics. Perhaps the strangest moment in the life of the of a theoretical physicist is that in which one realizes, all of a sudden, that one's life is being spent in pursuit of a kind of mystical experience that few of one's fellow humans share. I'm sure that most scientists of all kinds are inspired by a kind of worship of nature, but what makes theoretical physicists peculiar is that our sense of connection with nature has nothing to do with any direct encounter with it.[64]

The words of Aldous Huxley's final essay, *Literature and Science*, also retain their distinct level of insight: Literature is not concerned 'with regularities and explanatory laws, but with descriptions of appearances and the discerned qualities of objects perceived as wholes, with judgments, comparison and discriminations.'[65] Almost as an antidote for any obscuration of beauty, Goethe wrote in the middle of his life:

> We talk far too much. We should talk less and draw more. I personally should like to renounce speech altogether and, like organic Nature, communicate everything I have to say in sketches. That fig tree, this little snake, the cocoon on my window sill quietly awaiting its future – all these are momentous signatures. A person able to decipher their meaning properly would soon be able to dispense with the written or the spoken word altogether. The more I think of it, there is something futile, mediocre, even (I am tempted to say) foppish about speech. By contrast, how the gravity of Nature and her silence startle you, when you stand face to face with her, undistracted, before a barren ridge or in the desolation of the ancient hills.[66]

And, in the spirit of the Karamazov brothers, Huxley wrote the following stunning passage in *Antic Hay*, which captures the above feelings of Schlesinger and Goethe:

> "There are quiet places also in the mind," he said meditatively. "But we build bandstands and factories on them. Deliberately – to put a stop to the quietness. We don't like the quietness. All the thoughts, all the preoccupations in my head – round and round, continually.' He made a circular motion with his hand. 'And the jazz bands, the music-hall songs, the boys shouting the news. What's it for? What's it for? To put an end to the quiet, to break it up and disperse it, to pretend at any cost it isn't there. Ah, but it is; it is there, in spite of everything, at the back of everything. Lying awake at night, sometimes – not restlessly, but serenely, waiting for sleep – the quiet re-establishes itself, piece by piece; all the broken bits, all the fragments of it we've been so busily dispersing all day long. It re-establishes itself, an inward quiet, like this outward quiet of grass and trees. It fills one, it grows – a crystal quiet, a growing, expanding crystal. It grows, it becomes more perfect; it is beautiful and terrifying, yes, terrifying as well as beautiful. For one's alone in the crystal and there's no support from outside, there's nothing external and important, nothing external and trivial to pull oneself up by or to stand on, superiorly, contemptuously, so that one can look down. There's nothing to laugh at or feel enthusiastic about. But the quiet grows and grows. Beautifully and unbearably. And at last you are conscious of something approaching; it is almost a faint sound of footsteps. Something inexpressibly lovely and wonderful advances through the crystal, nearer, nearer. And, oh, inexpressibly terrifying. For if it were to touch you, if it were to seize and engulf you, you'd die; all the regular, habitual, daily part of you would die. There would be an end of bandstands and whizzing factories, and one would have to begin living arduously in the quiet, arduously in some strange, unheard-of manner. Nearer, nearer come the steps; but one can't face the advancing thing. One daren't."[67]

Graham Swift's novel *Waterland* develops this sentiment further: '[T]here's no saying what consequences we won't risk, what reactions to our actions, what repercussions, what brick towers built to be knocked down, what chasings of our own tales, what chaos we won't assent to in order to assure ourselves that, none the less, things are happening.'[68] In his *Private Thoughts*, Descartes meditates over the following similar ideas: 'It might seem strange that opinions of weight are found in the works of poets rather than philosophers. The reason is that poets wrote through enthusiasm and imagination; there are in us seeds of knowledge, as of fire in a flint; philosophers extract them by way of reason, but poets strike them out by imagination, and then they shine more bright.'[69] Metaphor also structures much of our folk science, with Empedocles calling the sea 'the sweat of the earth.'[70] Thales also believed the earth to be a disk floating on water, while Anaximenes thought the stars were nailed to a rotating sphere. With equal elegance, Lucretius demonstrates his folk science by inferring the behaviour of atoms through observing the movement of falling objects:

> When the atoms are travelling straight down through empty space by their own weight, at quite indeterminate times and places they *swerve* ever so little from their course, just so much that you can call it a change of direction. If it were not for this swerve, everything would fall downwards like rain-drops through the abyss of space. No collision would take place and no impact of atom on atom would be created. Thus nature would never have created anything. ... But the fact that the mind itself has no internal necessity to determine its every act and compel it to suffer in helpless passivity – this is due to the slight swerve of the atoms at no determinate time or place.[71]

Chomsky points out that 'whatever may be learned about folk science will have no relevance to the pursuit of naturalistic inquiry into the topics that folk science addresses in its own way, a conclusion taken to be a truism in a study of what is called 'the physical world' but considered controversial or false (on dubious grounds, I think) in the study of the mental aspects of the world.'[72] In his first Russell lecture, Chomsky commented:

> Surveying some recent experimental work, Monod remarks that there can be no doubt that animals are capable of classifying objects and relations according to abstract categories, especially geometric categories such as "triangle" and "circle". To some extent experimental work has even identified the neural basis for such analysis. This work suggests that there is a primitive, neurologically given analytic system which may degenerate if not stimulated at an appropriate critical period, but which otherwise provides a specific interpretation of experience, varying with the organism to some extent.[73]

In fact the human ability to categorise has been proven to exist in other animals: 'It was some time before the same underlying perceptual discontinuities were discovered in chinchillas and macaques, and even birds, leading to the opposite conclusion that the perceptual basis for categorical perception is a primitive vertebrate characteristic that evolved for general auditory processing, as opposed to specific speech processing.'[74] Most of the mechanisms involved in the language faculty exist in other species, leading many scientists to assume that 'the computational mechanism of recursion ... is recently evolved and unique to our species.'[75]

The interactions between Wernicke's area of the brain and Broca's area (the two central language centres, but not the 'language faculty' itself) reveal the avenues through which this experience takes place. Wernicke's area deals with nonverbal thoughts wishing to be expressed, including emotions, smells or images. Words begin to take shape in Broca's area, where phonemes are processed. When the individual intends to say a word, Broca's area then sends instructions to the facial motor cortex, at which point impulses are sent to the appropriate muscles in the mouth to articulate. Damage to Wernicke's area can result in the brain losing its grasp on semantic associations. Damage Broca's area, however, and 'though you might know what you want to say, you just can't send those impulses, effectively rendering you mute.' Duncan Graham-Rowe explains: 'When you listen to you inner voice, two things are happening. You "hear" yourself producing language in Wernicke's area as you construct it in Broca's area.' Using electrocorticography, neuroscientists have been able to map individual phonemes with neural signatures. Experiments by Eric Leuthardt discovered that spoken 'phonemes activated both the language areas and the motor cortex. But imagined speech – that inner voice – boosted the activity of the neurons in the Wernicke's area.' Leuthardt emphasises the importance of the internal dialogue to the way we process the world. His real-time reading of phonemes also implies that the technology will eventually be able to read an individual's mind – certain aspects of it, at least. But the library of phonemes is currently limited, and linguists 'have only a hazy understanding of the neurobiological processes that create language from thought.'[76] What we can be sure of, writes David Robson, is that 'Once early humans started speaking, there would be strong selection for mutations that improved this ability, such as the famous *FOXP2* gene, which enables the basal ganglia and the cerebellum to lay down the complex motor memories necessary for complex speech.'[77]

The conditions under which the innate mechanisms of the mind become activated will depend of the experience of the individual. Only on the occasion of sense (the words a child hears from its parents, in the case of language) will these principles become manifest. 'Knowledge,' then, is best understood as the growth and maturation of internal cognitive systems alongside the 'nurturing' effect of the environment. As Jackendoff puts it, 'just as the planets do not solve internalized differential equations in order to know where to go next, we might want to say that speakers do not invoke internalized formation rules and constraints in order to construct and understand sentences.'[78] In the case of abstract shapes, to take another example, we turn to Descartes:

> When first in infancy we see a triangular figure depicted on paper, this figure cannot show us how a real triangle ought to be conceived, in the way in which geometricians consider it, because the true triangle is contained in this figure, just as the statue of Mercury is contained in a rough block of wood. But because we already possess within us the idea of a true triangle, and it can be more easily conceived by our mind than the more complex figure of the triangle drawn on paper, we, therefore, when we see the composite figure, apprehend not it itself, but rather the authentic triangle.[79]

Around half a century later, Leibniz suggested that our senses are 'not sufficient to give us the whole of it, since the senses never give anything but instances, that is to say particular or individual truths.' He believed the 'necessary' truths of mathematics and geometry 'must have principles whose proof does not depend on instances, nor consequently on the testimony of the senses, although without the senses it would never have occurred to us to think of them.'[80] He later outlined the internal structures of the brain in his metaphor of a marble statue, which is chiseled to reveal its true nature, arguing that the mind plays a dominant role in the nature of its contents. It's in this spirit that biolinguistics searches for laws of form and internal structure which determine the growth and shape of cognitive faculties, viewing questions of 'function' and 'adaptation' as secondary. Bringing with him a conceptual revolution in naturalistic methodology, Galileo also felt that nature 'is written in the mathematical language, and the symbols are triangles, circles and other geometrical figures, without whose help it is humanly impossible to comprehend a single word of it, and without which one wonders in vain through a dark labyrinth.'[81] And midway through the twentieth century, Turing explained that similar principles apply to more general functions of the body:

> Many parts of a man's brain are the "centres" which control respiration, sneezing, following moving objects with the eyes, etc.: all the reflexes proper (not "conditioned") are due to the activities of these definite structures in the brain. Likewise the apparatus for the more elementary analysis of shapes and sounds probably comes into this category. But the more intellectual activities of the brain are too varied to be managed on this basis. The difference between the languages spoken on the two sides of the Channel is not due to differences in development of the

French-speaking and English-speaking parts of the brain. It is due to the linguistic parts having been subjected to different training.[82]

Colin McGinn discusses the limits of reason in the light of the innate, genetic endowments determining the mechanisms which enter into the interpretation of experience:

> since, as is commonly supposed, the genes work symbolically, by specifying programmes for generating organisms from the available raw materials, they must contain whatever information is necessary and sufficient for this feat of engineering. So, for example, they must somehow specify the structure and functioning of the heart, and they must supply rules for generating this organ from primitive biological components. The genes are, as it were, unconscious anatomists and physiologists, equipped with the lore pertaining thereto. But what goes for the body also goes for the mind: the genes must also contain the blueprint for constructing organisms with the (biologically based) mental properties those organisms instantiate. They must, then, represent the principles by which mental properties supervene on physical properties. They must, that is, specify instructions adequate for creating conscious states out of matter. And the same holds for other mental attributes: the genes 'know' how to construct organisms with intentionality, with personhood, with the capacity to make free choices, with rich systems of knowledge – just as they contain instructions for making organisms that embody innate universal grammar. This requires a grip on the natural principles that constitute these attributes, as well as mastery of the trick of engineering them from living tissue. The genes represent unconsciously what creationists ascribe to the mind of God. And since God has to know the answer to the philosophical problems surrounding these attributes, so too do the genes. In fact, they have known the answers for a very long time, well before we ever formulated the questions. ... My point here is just that if the representationality of the genes is admitted, then we can see that the limits of reason need not be the limits of all human epistemic systems. The genes really need to contain the information required to generate psychological organisms, this being their task in life, but conscious reason is under no particular obligation to recapitulate that achievement. So the genes have philosophical insight built into their very job description. Hence each cell in the human body "knows" more philosophy than shall ever be accessible to the frontal lobes.[83]

One might also ask in passing why most theists define their God as 'transcending' or going 'beyond' human senses, a divinity invisible and completely undetectable in any way. Anthony Simon Laden believes that this was certainly not a fault among the original worshippers, but rather, in attempting to see the invisible, our ancestors tried to pass beyond their physical limits. As with the biolinguistic stress on language's search for 'beauty,' he writes 'it is clear that the struggle to see in this way is just the struggle to move beyond the border of our ordinary experience and tap into something extraordinary.'[84] Laden notes a striking Feuerbachian connection between religious invisibility and two philosophical problems: alienation and skepticism: 'How does he know there isn't more out there than he is seeing? And if that is a possibility, then how does he even know that what he sees is what is out there, and not merely his own projection? In the fact of the terror that can come from unbridled skepticism, some will turn to faith, to an insistence that the invisible is there, whether or not it will reveal itself in plain sight.'[85] Religion is thus a highly effective cure of solipsism, while the solidarity of group worship removes the pain of alienation – both of which are founded on biological adaptations and not 'superadded' by an external force. The anarchist Josiah Warren seems to have been conscious of this:

> The blunder of our critic is in not knowing that our enterprise is not based on human inventions, but on Natural Laws, that are as old as the creation; and yet so new to most people's comprehension that the whole subject appears to them at first like a dream.[86]

The mythology of religion allowed past generations to escape from their material existence and create new worlds of metaphor and symbolism with their newly-acquired linguistic and imaginative capacities, depicting humans as separate from the rest of nature. But, as Lawrence warned, we should not allow our feet to sail too far from the ground:

> So it is: we all have our roots on earth. And it is our roots that now need a little attention, need the hard soil eased away from them, and softened so that a little fresh air can come to them, and they can breathe. For by pretending to have no roots, we have trodden the earth so hard over them they are starving and stifling below the soil. We have roots, and our roots are in the sensual, instinctive and intuitive body, and it is here we need fresh air of open consciousness.[87]

Thomas Bower, in *The Rational Infant*, adds that 'some developmental psychologists believe that children have certain innate cognitive structures concerning the spatial organisation of objects in three dimensions, reflected in their ability to make judgments of depth.' One specific innate ability has been demonstrated by 'a child which has never been exposed to heights' who is 'tested on the so-called "visual-cliff" ... to see whether it instinctively avoids the deep end,' which it does.[88] Peter Carruthers also argues that we have innate knowledge of our folk psychology, or 'a network of common-sense generalizations that hold independently of context or culture and concern the relationships of mental states to one another, to the environment and states of the body and to behavior.' This understanding includes beliefs that 'pains tend to be caused by injury, that pains tend to prevent us from concentrating on tasks, and that perceptions are generally caused by the appropriate state of the environment.'[89] Carruthers writes that the problem of 'the child's acquisition of psychological generalizations cannot be solved, unless we suppose that much of folk-psychology is already innate, triggered locally by the child's experience of itself and others, rather than learned.'[90] This inborn knowledge, to Russell, 'is not merely knowledge about the constitution of our minds, but is applicable to whatever the world may contain, both what is mental and what is non-mental.'[91] As with the language faculty, experience serves to trigger, and not to form, these innate structures, which provide 'rich perspectives for interpreting the mind-independent world,' including 'Gestalt properties, cause-and-effect, sympathy of parts, concerns directed to a common end, psychic continuity, and other mentally imposed properties. Properties that the mind imposes on the world, physicists can't find them.'[92] '[T]he role of experience is only to cause the innate schematism to be activated.'[93] This 'perspectivist' view of the mind was even occasionally invoked by Quine, despite his behaviourist commitments: 'Our talk of external things, our very notion of things, is just a conceptual apparatus that helps us to foresee and control the triggering of our sensory receptors in the light of previous triggering of our sensory receptors.'[94] 'There can be no doubt,' wrote Kant, 'that all our knowledge begins with experience. ... But though all our knowledge begins with experience it does not follow that it all arises out of experience.'[95] The same path is taken by our moral faculties, as Friedrich Lange understood:

> Just as a man is born with legs, although he only later learns to walk, so on the same way he brings with him into the world the organ that is to distinguish right and wrong, and the development of his mind necessarily calls the function of this organ into exercise.[96]

These have proven to be some of the reasons for the inadequacy of behaviourism:

> I think that there is some significance in the ease and willingness with which modern thinking about man and society accepts the designation "behavioral science". No sane person has ever doubted that behavior provides much of the evidence for this study – all of the evidence, if we interpret "behavior" in a sufficiently loose sense. But the term 'behavioral science' suggests a not-so-subtle shift of emphasis toward the evidence itself and away from the deeper underlying principles and abstract mental structures that might be illuminated by the evidence of behavior. It is as if natural science were to be designated "the science of meter readings". What in fact would we expect of natural science in a culture that was satisfied to accept this designation for its activities?[97]

Chomsky also points out in *Powers and Propsects* that 'What had been the topic of inquiry – behaviour, texts, etc. – is not just data, with no privileged status, standing alongside any other data that might prove relevant for the investigation of the mind. Behaviour and texts are of no more intrinsic interest than, say, observations of electrical activity of the brain, which has become quite suggestive in recent years.'[98] The cognitive revolution which began after Chomsky's linguistic revolution of generative grammar in the 1950s was, for Pierre Jacob, 'responsible for the shift away from the study of human behavior towards the study of internal mental states and processes that may or not give rise to observable behavior.'[99] Newton, to his own disbelief, managed to alter the generally accepted scientific methodologies and views to ones in which the goal is not to seek final explanations but to find the best theory we can of observable data through experiments. Physics no longer seeks to make sense of phenomenology, contrary to Quine's view that the acceptance of the existence of theoretical entities (or 'posits') is required to make sense of the 'flux of our experience.'[100] After the time of Aristotle, the conception of 'matter' and 'form' were separated, and physics concentrated on matter and eliminated form. But during the seventeenth century, as will be explored in the next two chapters, 'form' made a return as an explanation of the cognitive structures which shape our interpretation of events, rather than what events are actually 'made of.' Adding to the claim that experience serves to trigger innate mental structures, David Lightfoot comments:

> The grammar is one subcomponent of the mind, which interacts with other cognitive capacities or modules (acting as an 'innate

cognoscitive power,' to use the phrase of the British Neo-Platonist Ralph Cudworth). Like the grammar, each of the other modules may develop in time and have distinct initial and mature states. So the visual system recognizes triangles, circles and squares through the structure of the circuits that filter and recompose the retinal image. Certain nerve cells respond only to a straight line sloping downward from left to right, other nerve cells to lines sloped in different directions. The range of angles that an individual neuron can register is set by the genetic program, but experience is needed to fix the precise orientation specificity.[101]

A great deal of obscurity also surrounds the question of how the mind completes the blanks of a visual scene, specifically scenes pertaining to action and emotion. What we might call 'Shakespeare's problem' was briefly touched upon in *Henry V*: 'Piece out our imperfections with your thoughts ... and make imaginary puissance.' An interesting approach to exploring this problem was taken by Marie-Nöelle Metz-Lutz and her colleagues, who used the fMRI scans of subjects watching a play to gain insight into the cognitive processes involved in Coleridge's 'suspension of disbelief' and Keats' 'negative capability.' The point at which such suspension occurred

> was defined as when the subject's brain response tallied with a passage in the script intended to elicit such a response. The brain regions that fired at those moments included to areas involved in processing language and, specifically, in understanding metaphor, denoting the power of language to capture a spectator's attention. Both regions are also involved in processes of social and aesthetic judgements, probably governing appreciation of the writing style, plot or characterization.[102]

Brain activity also fell in areas responsible for constructing awareness about the self and the external world: 'Without activity in these regions, an observer will take the fictionalized reality of the play at face value, despite the sensory perception of the state, set and actors. Such results point to complete absorption in a play as a sort of hypnotic state involving the temporary loss of self-reference, and a disconnection from immediate sensory information – a distant feeling of being "carried away."'[103] As Cosmides and Tooby put it: 'although fiction seems to be processed as surrogate experience, some psychological subsystems reliably react to it as if it were real, while others reliably do not. In particular, fictional worlds engage emotion systems while disengaging action systems (just as dreams do).'[104] Tennyson illustrates another analogue – the theory of mind and self awareness – in his poem 'In Memoriam':

> The baby new to earth and sky,
> What his tender palm is prest
> Against the circle of the breast,
> Has never thought that "this is I."[105]

In *Language, Consciousness, Culture,* Jackendoff observes a less poetic distinction between the words 'look' and 'see,' with the latter implying a theory of mind: 'On can *look around* without looking at anything in particular; that is, *look* does not require a second argument. By contrast, one cannot *see* without seeing something; that is, *see* does require a second argument. (Even in the intransitive sentence *I can see*, the implication is that I can see *something*.)'[106] 'See' also allows epistemological errors: 'John saw a man on a horse in the desert' can be used in the case of a hallucination (John 'saw' a man on a horse, even if there wasn't one in the desert). From these and other verbs, we see how 'theory of mind is woven into the human conceptual system, not as a module, but as a suite of functions within the conceptual repertoire.'[107] Laying his cards on the table, Jackendoff sees

> at least three major domains of thought that cry out for substantial support from a genetic basis. The first is the understanding of the physical world: the identification of objects, their spatial configurations with respect to each other, the events in which they take part and interact, and the opportunities (or affordances) they offer for action on and with them. The second is the understanding of the social world: the identification of persons, their social roles with respect to each other (including such issues as kinship, dominance, group membership, obligations, entitlements, and morals), and characterization of their beliefs and motivations (so-called 'theory of mind'). The third is a basic algebra of individuation, categorization, grouping, and decomposition that undergirds both the previous systems as well as many others.[108]

Notice, however, the functionalist undertones to his argument that concepts grant 'affordances' for interacting with the world for a particular purpose. Gopnik and Meltzoff have also examined the linguistic development of the theory of mind, commenting that 'the emergence of belief words like "know" and "think" during the fourth year of life, after "see", is well established. In this case ... changes in the children's spontaneous extensions of these terms parallel changes in their predictions and explanations. The developing theory of mind is apparent both in semantic change and in conceptual change.'[109] But Gleitman contends that 'the child's problem isn't the inability to think about thinking, but only to find the evidence that the sound/word *think* is the item that expresses the concept "think" in English: a mapping problem rather than a conceptual problem. It simply is harder to glean, by observation alone, that thinkers are thinking than that, say, jumpers are jumping.'[110]

An essay by Winand Dittrich begins that 'it has long been known that specialized mechanisms exist in the human visual system that, on the one hand, permit us to categorize and remember faces with remarkable accuracy, but that, on the other hand, necessarily involve heuristics that can lead to mistaken perceptions.' Pawan Sinha has discovered an illusion which 'works best with a black-and-white photograph of someone looking sideways: if a negative of the image is made, the face seems to look in the opposite direction! This illusion shows that we have a specialized, modular mechanism for the socially crucial task of detecting gaze direction, which works on the assumption that the dark area is the iris. The knowledge that the iris is light instead of dark in a negative makes little or no difference: the illusion is cognitively impenetrable.'[111]

Shifting the science of the mind from metaphysical to epistemological inquiries, Locke often made note of similar cognitive limits (some of which are based on 'the vulgar error of the senses,' for Friedrich Lange): 'It seems probably to me, that the simple Ideas we receive from Sensation and Reflection, are the Boundaries of our Thoughts; beyond which, the Mind, whatever efforts it may make, is not able to advance one jot.'[112] Another barrier to understanding was identified by Thomas Reid, who pointed out the ambiguity of the term 'idea' in western philosophy. Locke, for instance, defined an idea as 'whatsoever is the object of understanding when a man thinks.' But as the first generation of neuroscience has confirmed, ideas are processes, not 'things.' But even Locke acknowledged the vague, arbitrary nature of such terms like '*ideas*, notions, or whatever else you please to call them.'[113] E. J. Lowe adds that 'the language of "ideas" is meant *not* to talk about a class of *entities* with various sensible properties of colour and shape, and so forth, but rather to talk about the *modes* or *manner* in which experiencing subjects are sensibly affected by perceptible objects like tables and rocks. On this view, our recourse to *nouns* and *adjectives* in the language of ideas, instead of *verbs* and *adverbs*, is just due to an inconvenient syntactical legacy of Indo-European languages.'[114] A. D. Woozley noted in an edition of Locke's *Essay* how, with the English language being exceptionally dedicated to nominalisation, 'noun-proneness indeed is one of the chief occupational diseases of philosophers.'[115] James Underhill issues a similar warning: 'Indo-European languages have caused us to fall under the illusion of taking our words for things.'[116] This level of abstraction is perhaps best approached by an anonymous character in *Annie Hall*, who at a fancy party boasts of his entrepreneurial spirit: 'Right now it's only a notion, but I think I can get money to make it into a concept, and later turn it into an idea.' Or perhaps we should just take the advice Ivan Karamazov gave the Devil during one of his eloquent speeches, and 'stop philosophizing, you ass!'[117]

In one of Salinger's short stories, a child prodigy called Teddy, when asked what he would do to change the education system, responds:

> I'd first just assemble all the children together and ... I'd get them to empty out everything their parents and everybody ever told them. I mean even if their parents just told them an elephant's big I'd make them empty that out ... I wouldn't even tell them an elephant has a trunk. I might show them an elephant, if I had one handy, but I'd let them just walk up to the elephant not knowing anything more about it than the elephant knew about them. The same thing with grass, and other things. I wouldn't even tell them grass is green. Colors are only names. I mean if you tell them the grass is green, it makes them start expecting the grass to look a certain way – your way – instead of some other way that may just be as good, and maybe much better.[118]

Contemporary philosophy of language can learn a great deal from Teddy, with the externalist dogma of 'reference' certainly being something worth 'emptying out.' Defending common sense against the Ivory Tower, Descartes suggested that 'those who have never studied judge much more reliably and clearly about salient matters than those who have spent all their time in the Schools.'[119] Discussing Jacques Monod's work, Chomsky writes:

> The evolution of human cortical structures was influenced by the early acquisition of a linguistic capacity, so that articulated language "not only has permitted the evolution of culture, but has contributed in a decisive fashion to the physical evolution of man," and there is no paradox in supposing that "the linguistic capacity that reveals itself in the course of the epigenetic development of the brain is now part of "human nature," itself intimately associated with other aspects of cognitive function which may in fact have evolved in a specific way by virtue of the early use of language."[120]

C. J. Lumsden's views on the relationship between the brain's structure and the evolution of culture resonate with Monod's: 'The relationship resembles one of reciprocating interaction, in which culture is generated and shaped by biological imperatives, while the course of genetic evolution shifts in response to cultural innovations.' Epigenetic rules may 'predispose mental development to take certain specific directions in the presence of certain kinds of cultural information.'[121] As a result, commonsense views of the world (or folk science) emerge and spread genetically, since 'certain ways of thinking will shape the individual nervous system, structurally as well as functionally. The presence or absence of stimulation affects the number of synaptic contacts, strengthening some and eliminating others.'[122] The evolution of language has enabled humans to use an unlimited set of conceptual structures and rich perspectives to conceive and communicate to others issues of how to live together in complex social arrangements, and so, writes James McGilvray, 'the science of language might well provide the key to what is distinctive to our minds and nature, to making sense of why we have the distinctive mental capacities we do and, in turn, making sense of how we can create our various forms of social organization.'[123] Echoing Humboldt, Hinzen promotes the following Cartesian-rationalist picture:

> The point of language in its ordinary use, to put it somewhat drastically, is not to relate us to the external world, but actually to *free* our mind from the control of the external stimulus, from to talk about the world as it actually is, as opposed to how it might be, was or will be. The hallmark of human language use is its creativity, or the *lack* of a connection to the immediate physical context and the adaptive challenges it poses. As a consequence of that, humans alone may have a history: lacking language, all non-human animals are stuck in the here and now.[124]

Animals, as Dennett puts it, are 'stuck in a world of appearances, making the best they can of how things seem and seldom, if ever, worrying about whether how things seem is how they truly are.'[125] To use Wittgenstein's example, 'A dog believes his master is at the door. But can he also believe his master will come the day after tomorrow?'[126] The anthropologist Stewart Guthrie, who has 'documented the rampant anthropomorphism in the world's religions,' believes 'We are hard-wired to see human-like beings everywhere.'[127] This tendency was noticed by Xenophanes 2600 years ago, who pointed out that if donkeys believed in gods theirs would have four legs. A more recent conclusion was drawn from a study in 2004 by

Nicholas Epley and John Cacioppo at the University of Chicago, who 'asked a group of volunteers to think about their beliefs, other peoples' beliefs, and God's beliefs on issues like capital punishment, while the researchers viewed their brain activity with a functional MRI scanner.' The closest resemblance to the subjects' own beliefs arose when they thought about God's views, triggering 'virtually identical' brain activity.[128] The less we know about the beliefs of another, it appears, the more we project our own beliefs onto them.

Other parts of our 'belief-system' are embedded in language and set the framework for interpreting the world: 'When we identify and name an object, we tacitly assume that it will obey natural laws. It will not suddenly disappear, turn into something else, or behave in some other "unnatural" way; if it does, we might conclude that we have misidentified and misnamed it. It is no easy matter to determine how our beliefs about the world of objects relate to the assignment of meanings to expressions. Indeed, it has often been argued that no principled distinction can be drawn.'[129] In a similar way, we do not 'believe' in our visual system, or our language faculty, or in the workings of our intestines. We 'know,' in some sense, the workings of our mind.

This relationship between the mind and the mind-external world was one Russell dealt with in his early work. He suggested that philosophers 'have maintained that relations are the work of the mind, that things in themselves have no relations, but that the mind brings them together in one act of thought and thus produces the relations which it judges them to have.'[130] Belief-systems cannot be answered for solely on a linguistic basis, it is commonly held: 'Mental states,' for John Searle, 'have *intrinsic* intentionality, material objects in the world that are used to represent something has *derived* intentionality. The most important form of derived intentionality is in language and there is a special name in English for this form of intentionality. It is called "meaning" in one of the many senses of that word.'[131]

But the commonsense notions of 'belief' and 'meaning' may fall outside the limits of naturalistic inquiry, and we may have to rely on ethnoscience (the study of a culture's systems of classifying knowledge and the attempt 'to reconstitute what serves as science for others, their practices of looking after themselves and their bodies, their botanical knowledge, but also their forms of classification, of making connections, etc.'[132]) through novels, music and other means to understand these concepts fully. During an interview with James McGilvray, Chomsky suggests that a 'belief' is

> just a description of what I did [e.g. take out an umbrella after seeing rain]. There's no independent notion of belief, desire, or

cause that enters into this discussion. ... [W]e don't even know that there are such entities as beliefs and desires. In fact, plenty of languages don't even have those words.[133]

From this perspective, 'belief' is just another factive verb like 'know' or 'think' or 'learn.' The reason such notions are so prominent, and the reason why a science of beliefs is likely impossible, is because

> English happens to be a highly nominalising language. ... You can't just tell a story using "tensor" and "molecules" and then say, ok, we [made a science of them]. You have to say what they are, and what the theoretical framework is in which you embed them, and so on. In these [belief-desire] cases, it's just not done.[134]

We should be cautious about our beliefs and desires simply because they are primarily encoded in language. We talk and think as if 'desires' are in one part of the brain and 'beliefs' are in another. One might ask whether the sentences we hear or read differ in any way from a simple re-description of what we see happening in the world. The connection is quite direct, and the principles of belief which enter into the explanation of events are not present. The sentence 'Tom wants a sandwich' is merely a re-description of the occasion on which I discover, one way or another, that Tom wants a sandwich. The internal systems of belief and intention are not disclosed in language (perhaps there's nothing to them altogether) so we should hesitate before we trust the things we talk about. 'It might turn out,' writes Chomsky,

> that beliefs and desires are attributed to creatures (perhaps humans) on entirely different grounds, perhaps as a reflection of instinctive modes of interpretation determined by innate endowment (common-sense), and that such attributes are systematically made even when the agents observed are considered to be acting in utterly irrational ways, or driven by instinct in contexts in which the question of rationality does not arise. ... There is evidence that young children attribute beliefs and plans to others well before they have terms to describe this; and the same may be true of adults generally, though most languages, it is reported, do not have terms corresponding to the English "belief."[135]

Donald Davidson, perhaps the chief nominaliser in modern analytic philosophy, has frequently displayed his lack of a naturalistic methodology in favour of a mixture between semantic formalism and E-language notions of 'propositions' and other supposed mind-external entities: 'A creature that has concepts, and hence beliefs, will no doubt also have other attitudes towards propositional contents, such as intentions, perceptions, memories, desires, hopes, and the rest.'[136] But contrary to Davidson, the philosophy of biolinguistics stresses that 'language formalism should be built straight from empirical data – much like the double-helix model in biology – rather than adhering to theorems that logic or mathematics have derived within their own systems. ... [L]inguistic formalism should use concepts from set theory without being built from its theorems.'[137] An account of how humans manage inductive reasoning to form 'beliefs' has not been reached in the sciences, but the descriptions of Pierce set the problem on its proper rationalist foundations:

> One familiar with the innate modes of reaction of subhuman organisms can readily hypothesise that the a priori is due to hereditary differentiations of the central nervous system which have become characteristic of the species, producing hereditary dispositions to think in certain forms. ... In the case of animals, we find limitations specific to the forms of experience possible for them. We believe we can demonstrate the closest functional and probably genetic relationship between these animal a prioris and our human a priori. Contrary to Hume, we believe, just as did Kant, that a "pure" science of innate forms of human thought, independent of all experience, is possible.[138]

Hume did believe, however, that science (or 'natural philosophy'), 'if cultivated with care, and encouraged by the attention of the public, may carry its researcher further, and discover, at least in some degree, the secret springs and principles, by which the human mind is activated in its operations.'[139] But the study of mind and nature cannot even get to the starting line, to paraphrase Strawson, if the terms and domain of inquiry are not defined, as the next chapter will explore.

REFERENCES

[1] Yevgeny Zamyatin, *We*, trans. Mirra Ginsburg, http://leecworkshops.wikispaces.com/file/view/yevgeny-zamyatin-we.pdf.
[2] James Joyce, *Ulysses*, ed. Jeri Johnson (Oxford University Press, 2008), p. 178.
[3] Noam Chomsky, 'Chomsky, Noam,' *A Companion to the Philosophy of Mind*, ed. Samuel Guttenplan (Oxford: Blackwell, 1996), p. 156.
[4] Matt Ridley, *Nature via Nurture: Genes, Experience, and What Makes Us Human* (New York: Harper Collins, 2003).

[5] Colin McGinn, 'Imagination,' *The Oxford Handbook of Philosophy of Mind*, ed. Brian P. McLaughlin, Ansgar Beckermann and Sven Walter (Oxford University Press, 2009), p. 598.
[6] Kurt Koffka, *Principles of Gestalt Psychology* (New York: Harcourt, Brace & World, 1935).
[7] Ray Jackendoff, *Foundations of Language: Brain, Meaning, Grammar, Evolution* (Oxford University Press, 2002), p. 308.
[8] Ban van Fraasen, 'Science, Materialism, and False Consciousness,' *Warrant in Contemporary Epistemology: Essays in Honor of Plantinga's Theory of Knowledge* (Lanham, MD.: Rowman & Littlefield, 1996), p. 180.
[9] Diderot, lettre à Étienne Noël Damilaville, 3 novembre 1760, *Correspondence*, ed. G. Roth (Paris: Éditions de Minuit, 1955-70), vol. III, p. 216; trans, McGilchrist, in Iain McGilchrist, *The Master and His Emissary: The Divided Brain and the Making of the Western World* (Yale University Press, 2010), p. 506.
[10] John Locke, *An Essay Concerning Human Understanding*, ed. A. D. Woozley (Glasgow: William Collins Sons & Co Ltd, 1964, repr. 1984) p. 65 (emphasis his).
[11] *Charles Darwin's Notebooks 1836-44*, ed. P. H. Barrett, P. J. Gautrey, S. Herbert, D. Kohn and S. Smith (Cambridge University Press, 1987), D 26, M 84.
[12] Kate Douglas, 'Tickling a rat...?' *New Scientist*, 17 July 2010.
[13] Wolfram Hinzen, *An Essay on Names and Truth* (Oxford University Press, 2007), pp. 160-1.
[14] M. R. Bennett and P. M. S. Hacker, *Philosophical Foundations of Neuroscience* (Oxford: Blackwell, 2003), p. 1.
[15] Michael Lockwood, *Mind, Brain, and the Quantum* (Oxford: Blackwell, 1989), p. 159.
[16] Galen Strawson, *Real Materialism and Other Essays* (Oxford University Press, 2008), p. 257.
[17] Noam Chomsky, 'Comments: Galen Strawson, Mental Reality,' *Philosophy and Phenomenological Research*, 58(2), June 1998: 440 (437-41); Galen Strawson, *Mental Reality*, 2nd ed. (MIT Press, 1994/2010), p. 62.
[18] Piet Hut and Roger Shepard, 'Turning "the Hard Problem" upside down and sideways,' *Explaining Consciousness: The Hard Problem*, ed. Jonathan Shear (Massachusetts: MIT Press, 1997), pp. 305-22.
[19] David C. Geary, 'The evolution of general fluid intelligence,' *Foundations in Evolutionary Cognitive Neuroscience*, eds. Steven M. Platek and Todd K. Shackelford (Oxford University Press, 2009), p. 46 (22-56).
[20] François Jacob, *The Possible and the Actual* (University of Washington Press, 1982), p. 10.
[21] William G. Lycan, 'Chomsky on the Mind-Body Problem,' *Chomsky and His Critics*, eds. Louise M. Antony and Norbert Hornstein (Oxford: Blackwell, 2003), p. 24.
[22] Martin Heidegger, *Poetry, Language, Thought*, ed. J. Glenn Gray, trans. Albert Hofstadter (New York: Harper and Row, 1971), p. 12.
[23] Noam Chomsky, *Problems of Knowledge and Freedom: The Russell Lectures* (London: Barrie & Jenkins, 1972), pp. 45-6.
[24] Michael Tanner, *Schopenhauer* (London: Routledge, 1999), p. 6.
[25] Marx X. Wartofsky, *Conceptual Foundations of Scientific Thought: An Introduction to the Philosophy of Science* (New York: Macmillan, 1968), p. 29.
[26] David Hume, *A Treatise of Human Nature*, ed. L. A. Selby-Bigge, 2nd ed. (Oxford: Clarendon, 1739-40/1978), p. xv.
[27] Strawson, *Real Materialism and Other Essays*, p. 26.
[28] Werner Heisenberg, *Physics and Philosophy: The Revolution in Modern Science*, Lectures at the University of St Andrews, Winter 1955-6 (London: Penguin, 1989), p. 25.
[29] Cited in Paul McEvroy, *Niels Bohr: Reflections on Subject and Object* (San Francisco: MicroAnalytix: 2001), p. 291.

[30] Steven Pinker, *The Stuff of Thought: Language as a Window into Human Nature* (London: Penguin, 2008), p. 158.
[31] David Hume, *An Enquiry Concerning Human Understanding*, in *Enquiries Concerning the Human Understanding and Concerning the Principles of Morals*, ed. L. A. Selby-Bigge, 2nd ed. (Oxford: Clarendon Press, 1902), p. 73 (emphasis his).
[32] David Hume, *Correspondence*, ed. H. W. Turnball (Cambridge University Press, 1959-77/1692-3), p. vol. 3, p. 240.
[33] Hume, *An Enquiry Concerning Human Understanding*, p. 76 (emphasis his).
[34] Friedrich Albert Lange, *The History of Materialism and Criticism of its Present Importance*, trans. Ernest Chester Thomas, 3rd ed. (London: Routledge & Kegan Paul, 1957), p. 164 (emphasis his).
[35] Ibid., pp. 361-2.
[36] Pinker, *The Stuff of Thought*. pp. 223-4.
[37] Jessica Griggs, 'To be quantum is to be uncertain,' *New Scientist*, 23 June 2012.
[38] Pinker, *The Stuff of Thought*, pp. 257-8.
[39] Edward Abbey, *A Voice Crying in the Wilderness (Vox Clamantis in Deserto): Notes from a Secret Journal*, (St. Martin's Press, 1990).
[40] Lucretius, *One the Nature of the Universe*, trans. Ronald Melville (Oxford University Press, 2008), Book I, l.915-920, p. 29.
[41] Joseph Priestley, *Priestley's Writings on Philosophy, Science, and Politics*, ed. John A. Passmore (New York: Collier Books, 1965), p. 55.
[42] Graham Lawton, 'The Grand Delusion,' *New Scientist*, 14 May 2011.
[43] Jeffrey Gray, *Consciousness: Creeping Up on the Hard Problem* (Oxford University Press, 2004), p. 311.
[44] Turing, 'Can Digital Computers Think?', *The Essential Turing*, ed. B. Jack Copeland (Oxford: Clarendon Press, 2004), p. 484.
[45] Leo Tolstoy, *War and Peace*, trans. Louise and Aylmer Maude, ed. Henry Gifford (Oxford University Press, 1991, repr. 2008), p. 1292.
[46] David Hume, *An Enquiry Concerning Human Understanding*, ed. L. A. Selby-Bigge, 3rd ed. rev. by P. H. Nidditch (Oxford University Press, 1748/1975), Section VII, Part I, §57, p. 72
[47] Semir Zeki, *A Vision of the Brain* (Oxford: Blackwell, 1993).
[48] K. O'Regan and A. Noë, 'A sensorimotor account of vision and visual consciousness,' *Behavioral and Brain Sciences*, 24, 2001: 939-1031.
[49] Wolfram Hinzen, *Mind Design and Minimal Syntax* (Oxford University Press, 2006), p. 65.
[50] Gray, *Consciousness*, p. 35.
[51] Terrence Sejnowski, 'The product of our neurons,' *New Scientist*, 4 February 2012.
[52] McGilchrist, *The Master and His Emissary*, p. 21.
[53] John Dupré, 'How the Mind Works by Steven Pinker,' *Philosophy of Science* 66(3), September 1999: 489 (489-93).
[54] Anthony Gottlieb, 'Neurons v free will,' *Intelligent Life*, March/April 2012.
[55] Friedrich Nietzsche, *Beyond Good and Evil*, 1886, §17.
[56] Noam Chomsky, *The Science of Language: Interviews with James McGilvray* (Cambridge University Press, 2012), p. 280.
[57] Strawson, *Real Materialism and Other Essays*, p. 246. Citing Donald Davidson, 'Agency,' in *Agent, Action, and Reason*, ed. R. Binkley, R. Bronaugh and A. Marras (Toronto: University of Toronto Press, 1971), p. 23.
[58] Gray, *Consciousness*, p. 9.
[59] Cited in Susan Blackmore, *Conversations on Consciousness* (Oxford University Press, 2005), p. 153 (emphasis his).
[60] Russell, 'Introduction: Materialism, Past and Present,' in Lange, *The History of Materialism and Criticism of its Present Importance*, p. xvi.

[61] Noam Chomsky, *New Horizons in the Study of Language and Mind* (Cambridge University Press, 2000), p. 77.
[62] Percy Bysshe Shelley, 'To the Moon.'
[63] Colin McGinn, *Shakespeare's Philosophy* (New York: Harper Collins, 2006), p. 166.
[64] Lee Smolin, *The Life of the Cosmos* (London: Phoenix, 1997).
[65] Aldous Huxley, *Literature and Science* (London: Chatto & Windus, 1963), p. 10.
[66] Cited in Aldous Huxley, *The Doors of Perception and Heaven and Hell* (London: Flamingo, 1994 [1954]), p. 51.
[67] Aldous Huxley, *Antic Hay* (Dalkey Archive Press, 1997).
[68] Graham Swift, *Waterland* (London: William Heinemann, 1983), p. 47.
[69][69] Cited in Harold Bloom, *The Anxiety of Influence: A Theory of Poetry*, 2nd ed. (Oxford University Press, 1997), p. 40.
[70] Cited in Derek Gjersten, *Science and Philosophy: Past and Present* (Oxford University Press, 1989), p. 14.
[71] Cited in ibid., p. 44.
[72] Chomsky, *New Horizons in the Study of Language and Mind*, p. 199.
[73] Chomsky, *Problems of Knowledge and Freedom*, pp. 19-20.
[74] Marc. D. Hauser, Noam Chomsky, W. Tecumseh Fitch, 'The Faculty of Language: What Is It, Who Has It, and How Did It Evolve?', *Science*, 298(5598), 2002: 1572.
[75] Ibid., 1573.
[76] Duncan Graham-Rowe, 'Mind readers,' *New Scientist*, 28 May 2011.
[77] David Robson, 'A brief history of the brain,' *New Scientist*, 24 September 2011.
[78] Jackendoff, *Foundations of Language*, p. 56.
[79] Cited in Noam Chomsky, 'Recent contributions to the theory of innate ideas,' *Synthese*, 17(1), March 1967 (Dordrecht, Holland: D. Reidel Publishing Company): 10.
[80] Gottfried Leibniz, *New Essays on Human Understanding* (c.1704), in *Leibniz: Philosophical Writings*, ed. G.H.R. Parkinson, trans. Mary Morris and G. H. R. Parkinson, (London: J.M. Dent & Sons, 1973), Preface, pp. 150-1.
[81] Galileo Galilei, *The Assayer* (University of Pennsylvania Press, 1960/1623).
[82] Turing, 'Intelligent Machinery,' *The Essential Turing*, p. 423.
[83] Colin McGinn, 'The Problem of Philosophy,' www.nyu.edu/gsas/dept/philo/courses/consciousness97/papers/ProblemOfPhilosophy.html.
[84] Anthony Simon Laden, 'Transcendence without God: On Atheism and Invisibility,' *Philosophers without Gods: Meditations on Atheism and the Secular Life*, ed. Louise M. Antony (Oxford University Press, 2007), p. 126.
[85] Ibid., p. 130.
[86] Josiah Warren, *Practical Applications of the Elementary Principles of 'True Civilization,'* (1873), in Marshall S. Shatz (ed.), *The Essential Works of Anarchism* (New York: Quadrangle Books, 1972), p. 449.
[87] *The Complete Poems of D. H. Lawrence*, ed. Vivian de Sola Pinto and Warren Roberts (Harmondsworth: Penguin, 1977), p. 418.
[88] T. G. R. Bower, *The Rational Infant* (New York: W. H. Freedman, 1989), cited in E. J. Lowe, *Routledge Philosophy Guidebook to Locke on Human Understanding* (Routledge, T. J. Press Ltd, Cornwall, 1995), p. 30.
[89] 'Rationalism vs. Empiricism,' Stanford Encyclopedia of Philosophy, http://plato.stanford.edu/entries/rationalism-empiricism.
[90] Peter Carruthers, *Human Knowledge and Human Nature* (Oxford University Press, 1992), p. 121.
[91] Bertrand Russell, *The Problems of Philosophy* (Oxford University Press, 1959/1912), p. 50.
[92] Noam Chomsky, 'Language and the Cognitive Science Revolution(s),' lecture given at Carleton University, 8 April 2011, http://chomsky.info/talks/20110408.htm.

[93] Chomsky, 'Recent contributions to the theory of innate ideas,' 9.
[94] W. V. O. Quine, *Theories and Things* (Cambridge, MA.: Belknap Press of Harvard University Press, 1981), p. 1.
[95] Immanuel Kant, *Critique of Pure Reason*, trans. Norman Kemp Smith (London: St Martin's Press, 1963), p. 41.
[96] Lange, *The History of Materialism and Criticism of its Present Importance*, p. 20.
[97] Noam Chomsky, *Language and Mind*, 3rd ed. (Cambridge University Press, 2006), p. 57.
[98] Noam Chomsky, *Powers and Prospects: Reflections on Human Nature and the Social Order* (London: Pluto, 1996), p. 12.
[99] Pierre Jacob, 'Chomsky, Cognitive Science, Naturalism and Internalism,' 2001, http://hal.inria.fr/docs/00/05/32/33/PDF/ijn_00000027_00.pdf, p. 4.
[100] Cited in Cory Juhl and Eric Loomis, *Analyticity* (London: Routledge, 2010), p. 131.
[101] David Lightfoot, 'Plato's Problem, UG, and the language organ,' *The Cambridge Companion to Chomsky*, ed. James McGilvray (Cambridge University Press, 2005), p. 52.
[102] Giovanni Frazzetto, 'Powerful acts,' *Nature*, 482(7386), 12 February 2012: 467.
[103] Ibid.
[104] Leda Cosmides and John Tooby, 'Does Beauty Build Adapted Minds? Toward an EvolutionaryTheory of Aesthetics, Fiction, and the Arts,' *SubStance*, 30, 2001: 8 (6-27).
[105] Alfred Tennyson, *The Major Works*, ed. Adam Roberts (Oxford University Press, 2009), p. 230.
[106] Ray Jackendoff, *Language, Consciousness, Culture: Essays on Mental Structure* (Massachusetts: MIT Press, 2007), p. 205.
[107] Ibid., p. 216.
[108] Ray Jackendoff, *Foundations of Language: Brain, Meaning, Grammar, Evolution* (Oxford University Press, 2002), pp. 274-5.
[109] Alison Gopnik and Andrew Meltzoff, *Words, Thoughts and Themes* (MIT Press, 1997), p. 121.
[110] Lila Gleitman, 'Round Table: Language Universals: Yesterday, Today, and Tomorrow,' *Of Minds and Language: A Dialogue with Noam Chomsky in the Basque Country*, ed. Massimo Piatello-Palmarini, Juan Uriagereka and Pello Salaburu (Oxford University Press, 2009), p. 248.
[111] Winand Dittrich, 'When bottom-up meets top-down,' *Trends in Cognitive Sciences*, 5(4), April 2001: 137.
[112] Lange, *The History of Materialism and Criticism of its Present Importance*, p. 3; Locke, *An Essay Concerning Human Understanding*, ed. Roger Woolhouse (London: Penguin, 1997), p. 282.
[113] Locke, *An Essay Concerning Human Understanding*, p. 64 (emphasis his).
[114] Lowe, *Routledge Philosophy Guidebook to Locke on Human Understanding*, p. 45 (emphasis his).
[115] Locke, *An Essay Concerning Human Understanding*, p. 32.
[116] Underhill, *Humboldt. Worldview and Language*, p. 43.
[117] Fyodor Dostoevsky, *The Brothers Karamazov*, trans. Richard Pevear and Larissa Volokhonsky (London: Vintage, 1992), p. 641.
[118] J. D. Salinger, 'Teddy,' in Marjorie McCorquodale, 'Poets and Scientists,' *Bulletin of the Atomic Scientists*, November 1965, p. 19.
[119] René Descartes, *Rules for the Direction of the Mind*, in *The Philosophical Writings of Descartes*, trans. J. Cottingham, R. Stoothoff, D. Murdoch and A. Kenny (Cambridge University Press, 1985/1618-28), vol. 1, p. 16.
[120] Chomsky, *Problems of Knowledge and Freedom*, p. 10.

[121] C. J. Lumsden, 'Gene-culture coevolution: culture and biology in Darwinian perspective,' in D. de Kerckhove and C. J. Lumsden (eds.) *The Alphabet and the Brain: The Lateralization of Writing* (Berlin: Springer-Verlag, 1988), p. 17, 20.
[122] McGilchrist, *The Master and His Emissary*, p. 246.
[123] James McGilvray, 'Introduction,' *The Cambridge Companion to Chomsky*, p. 9.
[124] Hinzen, *Mind Design and Minimal Syntax*, p. xiii.
[125] Daniel Dennett, *Freedom Evolves* (London: Penguin, 2003).
[126] Wittgenstein, *Philosophical Investigations*, §174.
[127] Cited in Douglas Fox, 'In our own image,' *New Scientist*, 27 November 2010.
[128] Fox, *New Scientist*, 27 November 2010, see *Proceedings of the National Academy of Sciences*, vol. 16, p. 2513.
[129] Chomsky, *Rules and Representations* (Columbia University Press, 1980), p. 225.
[130] Russell, *The Problems of Philosophy*, p. 51.
[131] John Searle, in Guttenplan, *A Companion to the Philosophy of Mind*, p. 386.
[132] Marc Augé, *The War of Dreams: Exercises in Ethno-Fiction* (London: Pluto Press, 1999), p. 118.
[133] Chomsky, *The Science of Language*, p. 138.
[134] Ibid., p. 139.
[135] Chomsky, *New Horizons in the Study of Language and Mind*, p. 91, 119.
[136] Davidson, 'The Perils and Pleasures of Interpretation,' *The Oxford Handbook of Philosophy of Language*, ed. Ernest Lepore and Barry C. Smith (Oxford: Clarendon, 2006), p. 1065 (emphasis his).
[137] Tom Roeper, 'The Acquisition of Recursion: How Formalism Articulates the Child's Path,' *Biolinguistics*, 5(1-2), 2011: 59, n. 1 (57-86).
[138] Cited in Chomsky, *Language and Mind*, p. 84.
[139] Hume, *An Enquiry Concerning Human Understanding*, Section I, 'Of the different Species of Philosophy,' p. 14.

4

THE MENTAL AND THE NATURAL

Ontological questions are generally beside the point, hardly more than a form of harassment.
                                                        NOAM CHOMSKY[1]

When we ask, with Locke, whether matter can think, that is just as if we were to ask whether matter can show the time.
                                                        FRIEDRICH LANGE[2]

Studying metaphor, Andrew Goatly tells us, is important because it 'demonstrates, in an exaggerated way, how all linguistic classification constructs a representation of experience on the basis of selective perception and selective ignoring of aspects of the world.'[3] There are many examples of the conceptual metaphors English is riddled with, including ANGER IS A FLUID IN A CONTAINER (he boiled with rage); LOVE IS NUTRITION/FOOD (it could be essential to our understanding of emotion and love that our metaphors evoke images of motion – or, as Kundera put it: 'I have said before that metaphors are dangerous. Love begins with a metaphor. Which is to say love begins at the point when a woman enters her first word into our poetic memory.'[4]); TIME IS MONEY; IDEAS ARE SUBSTANCES (drop a hint, pass it on, share a joke, give a promise before taking it back, gather information to grasp an idea, and hold a thought before toying with it); THE MIND IS A CONTAINER; LANDSCAPES ARE BODIES (face of the earth, arm of the land, backbone, brow, foot of a mountain, mouth of a river, fringe of a forest, heart of a jungle); COMMERCE IS A STREAM (the Exchequer must 'regulate its flow and protect it from obstructions'[5]); GOOD IS UP, BAD IS DOWN; EMOTIONS ARE LIQUID (they run high, rise, subside, upset, pour out, dry, flood) and so on.

> Even the little grammatical words have a physical provenance. Sometimes it is evident in modern English, as in the pronoun *it* (A SITUATION IS A THING) and the prepositions *in* (TIME IS SPACE),

to (INTENTION IS MOTION TOWARD A GOAL), and *among* (AFFILIATION IS PROXIMITY).[6]

We also conceptualise much of our experience in terms of 'containers,' from when we get 'out' of bed to when we walk 'in' to work (see the example of 'brown house' below). Problems are thought of as physical obstacles to be overcome (barriers, walls, hurdles etc.). It also follows that 'if our conceptual categories are based on universal infant experiences we share conceptual systems with our fellow humans.'[7] Pinker holds the view that these conceptual metaphors form the basis of most Western philosophy, since

> There are only competing metaphors, which are more or less apt for the purposes of the people who live by them. ... Descartes's philosophy is based on [the conceptual metaphor] KNOWING IS SEEING, Locke's on THE MIND IS A CONTAINER, Kant's one MORALITY IS A STRICT FATHER, and so on. It is a creation of the human body senses, growing out of the activities of moving along a path and of collecting, constructing, and measuring objects.[8]

Taking this into account, 'there seems to be no obvious reason why the progress of the language-game we are playing should have anything in particular to do with the way the rest of the world is.'[9] Goatly continues by pointing out that 'we have difficulty conceptualizing smells, because our language is poor in words to label them. This inadequacy of the conceptual system is the very reason that when we encounter them they evoke images, with all their associations. Trying to identify a familiar but uncategorized smell or taste, we explore our episodic memory for the events and places in which we previously encountered it, and with which we associate it.'[10] Peter Stockwell, in his introduction to the new (but hopeless and misguided) field of 'cognitive poetics,' writes that 'all of our experiences, knowledge, beliefs and wishes are involved in and expressible only through patterns of language that have their roots in our material existence.'[11] But, as with Lakoff, he takes a pragmatist view of metaphor, arguing that metaphors are primarily tools suited to humans, dismissing the question of how such tools emerged to begin with. And, as with the moral relativists, the 'embodied' scholars like Goatly presuppose the existence of logical necessity and truth in order to even discuss the moulding effects of conceptual metaphors: 'Even if we grant Lakoff the point that abstract concepts are somehow metaphorical, the crucial next step is to show how thinking metaphorically can be rational, not to abandon rationality altogether.'[12]

McGilchrist also falls prey to a crypto-pragmatist conception of truth in arguing that the 'problem' of God

> is not reducible to a question of a factual answer to the question "does God exist?", assuming for the moment the expression "a factual answer" has a meaning. It is having an attitude, holding a disposition towards the world, whereby that world, as it comes into being for me, is one in which God belongs. The belief alters the world, but also alters me. It is true that God exists? Truth is a disposition, one of being true to someone or something. One cannot believe in nothing and thus avoid belief altogether, simply because one cannot have *no* disposition towards the world, that being in itself a disposition. Some people choose to believe in materialism; they act 'as if' such a philosophy were true. An answer to the question whether God exists could only come from my acting "as if" God is, and in this way being true to God, and experiencing God (or not, as the case may be) as true to me. It I am a believer, I have to believe in God, and God, if he exists, has to believe in me.[13]

In *The Poetics of Mind*, Raymond Gibbs explained that we 'conceptualize their experiences in figurative terms via metaphor, metonymy, irony, oxymoron, and so on, and these principles underlie the way we think, reason, and imagine.'[14] It's been repeatedly shown that those with absolutely no knowledge of language still believe that 'kiki' suggests a spike-shaped object, whilst 'bouba' suggests a softly rounded bulbous object, suggesting that certain elements of cognition operate independently of language and are based on forms of synaesthesia which arise from the neural cross-wiring of certain regions of the brain.[15]

English also employs the metaphor of THE FUTURE IS AHEAD OR IN FRONT in various ways. This is even true to the extent that Lynden Miles has demonstrated how English speakers unconsciously sway forwards when thinking and speaking about the future. It's also been discovered that the Kuuk thaayorre language has no words for 'left' or 'right,' but relies purely on cardinal directions, perhaps shaping, in some way, its speaker's spatial cognition. Though we should be wary of treading where Sapir and Whorf did, Barbara Landau has recently called language a 'momentary mechanism' which causes momentary effects of modulating or directing attention, lasting seconds and not lifetimes.[16] Reviewing recent work on the representation of temporal events by birds, Randy Gallistel also summarises that 'the remembered past of the bird is temporally organized just as in our own. The birds compute elapsed time intervals and compare them to other intervals in memory. ... Like us, birds reason about time.'[17] Drawing on experiments which suggests that jays 'selectively re-cache the food of others observed them cache' when they are no longer being watched, Gallistel also suggests that 'nonverbal animals represent the likely intentions of others and reason from their own actions to the likely future actions of others.'[18]

It's also been shown how 'we are more likely to remember an object if we know its name, which may explain why we have so few memories of early childhood.'[19] The edge detection component of object perception may emerge from prenatal development, with other major components, such as perception of objects over occlusion, developing postnatally.[20] We don't think of the struggles of life in terms of putting on socks, but in terms of a 'journey,' with 'physical objects' to overcome, with different routes, guides, crossroads, landmarks and so on. In fact even the word 'literally' is used figuratively ('I literally died of exhaustion'). It's hard to define what 'literal' language use actually is, outside of technical terminology. In various experiments it's been shown that idioms and indirect speech acts ('Could you pass me the sugar?') 'took *no* longer to understand than either literal sentences or direct requests when these sentences were read inappropriate contexts.'[21] This has social advantages, as 'tag questions,' 'hedges' and the infamous 'Australian questioning intonation' ensure we don't come across as too assertive. Walt Whitman noticed that 'Slang, or indirection, is an attempt of common humanity to escape from bald literalism, and express itself illimitably, which in highest walks produces poets and poems.'[22] Here is Rorty's view:

> Tossing a metaphor into a conversation is like suddenly breaking off the conversation long enough to make a face, or pulling a photograph out of your pocket and displaying it, or pointing at a feature of the surroundings, or slapping your interlocutor's face, or kissing him. Tossing a metaphor into a text is like using italics, or illustrations, or odd punctuations or formats. All these are ways of producing effects on your interlocutor or your reader, but not ways of conveying a message.[23]

Rorty's pragmatist credentials commit him to the position that metaphors have no semantic content in themselves (contrary to the widespread existence of conceptual metaphors structuring much of cognition), since they are not propositional (where is the truth-condition in 'Juliet is the sun?'). Richard Lederer believes that 'you have to marvel at the unique lunacy of the English language ... in which your house can simultaneously burn up and burn down and your car can slow up and burn down, in which you fill in a form by filling out a form, in which your alarm clock goes off by going on ... and in which you first chop a tree down – and then you chop it up.'[24] 'Things that we claim are *underwater* and *underground* are obviously surrounded by, not under the water or ground.'[25] And why do we call it 'after dark' when it is 'after light'? There are countless other examples, and some may seem trivial, but they all reflect deep mental models which we use to make sense of our lives. As the 'father of modern anthropology' Franz Boas put it, 'the categories of language compel us to see the world arranged in certain definite conceptual groups which, on account of our lack of knowledge of linguistic processes, are taken as objective categories and which, therefore, impose themselves upon the form of our thoughts.'[26]

Language itself has its own physical laws, namely; a concept of space in our prepositions; a concept of matter in our nouns; a concept of time in our tenses; and a concept of causality in our verbs. Our language also provides four frames in which to interpret 'matter': countable things (an apple); masses (much applesauce); plurals (many apples); and collections (a dozen apples). All of this amounts to an intuitive theory of the physical world, one which becomes contradictory and counterintuitive only when we inspect our language and frameworks through which it operates. The conceptual frameworks of the mind force a certain interpretation of the world, which Da Vinci thought was based on 'simple and plain experience which is the true mistress.'[27]

Using our imaginative capacities, primarily of the brain's right hemisphere, concepts form independent of language, though language helps organise and manage our concepts as the biolinguistic program stresses. There have been a number of studies in the last decade which have shown how categorical perception is not unique to humans, since any animal living in an environment surrounded by certain types of food, prey or shelter needs to form categories of them in order to survive.[28] Russell believed that all of us are 'guilty, unconsciously and in spite of explicit disavowals, of a confusion in [our] imaginative picture of matter.'[29] Markus Arndt qualifies that assumption, who states the possibility that the fundamental 'matter' of the universe are neither waves nor particles. If this is correct, then 'waves' and 'particles' are 'constructs of our mind to facilitate everyday talking.'[30] Lera Boroditsky, writing in *Scientific American*, has stated:

> Studies have shown that changing how people talk changes how they think. Teaching people new color words, for instance, changes their ability to discriminate colors. And teaching people a new way of talking about time gives them a new way of

thinking about it. ... People rely on language even when doing simple things like distinguishing patches of color, counting dots on a screen or orienting in a small room.[31]

Priestley seems to have adopted this marginally (very marginally) Whorfian sentiment when concluding that 'Words are of great use in the business of thinking, but are not necessary to it.'[32] Before certain absurd questions of artificial intelligence arose over the latter half of the twentieth century (grounded as they were in the work of artisans like Jacques de Vaucanson, who constructed artefacts which imitated animal behaviour), the mathematician Alan Turing wrote in an often misinterpreted paper: 'The original question, "Can machines think?" I believe to be too meaningless to deserve discussion.'[33] All organisms interpret the world in different ways, and so what's a mystery for a dog is not a mystery for a human. The computation in the mind of a dog may be less sophisticated than the one of a human when they each watch a plane move through the sky, but the fact is we simply don't know enough about 'thought' in order to reserve the term only for ourselves. Asking whether any other animal (or even an infant) is 'thinking' is like asking whether an airplane is 'flying.' Like the concept of 'intelligence,' writes John Duncan, these 'are strictly questions of terminology, not content – questions not about how the world is but about what we will call things. "Intelligence" is simply a word from everyday language. ... [W]e have every reason to suspect that it will have no "real essence" at all. ... Accordingly, we are free to call emotion or memory a part of 'intelligence,' or not to.'[34]

One of Jackendoff's central objections to what he regards as the 'syntactocentric' biolinguistics enterprise is that 'there is a curious tension between stressing that language is the means for the free expression of thought, following the Cartesians, and formalising language in a way that seems to deny thought any independent status.'[35] But biolinguistics does not claim with the German philologists that 'thought' is entirely dependent on language (hence the modular approach). Rather, it simply seeks to give a naturalistic account of linguistic aspects of thought. To borrow Priestley's phrase, it studies those aspects of thought 'termed linguistic.' These modular approaches are similar to Hume's view of philosophy, the central of role of which being to find 'the different operations of the mind, to separate them from each other, to class them under their proper heads.'[36]

Nevertheless, 'The whole thinking process,' noted Turing in 1951, 'is still rather mysterious to us.'[37] Almost a year later he refined his opinion, confessing 'I don't want to give a definition of thinking, but if I had to I should probably be unable to say anything about it than that it was a sort of buzzing that went on inside my head.'[38] These views were given shortly before his tragic death of effectively being murdered by the British state for the unspeakable sin of homosexuality by having female hormones injected into his bloodstream, which forced his body to eventually grow breasts and his mind to descend into chronic depression – an official apology for which David Cameron recently rejected giving, continuing his lifelong search for justice and truth. In a free society this kind of scandal would cause an uproar in the media and intellectual circles generally, as opposed to the virtual silence of the corporate media.

Turing had also written in 'Intelligent Machinery' that 'intellectual activity consists mainly of various kinds of search [computation].'[39] He speculated that a reason many reject the notion that machines can 'think' stems from an 'unwillingness to admit the possibility that mankind can have any rivals in intellectual power. This occurs as much amongst intellectual people as amongst others: they have more to lose. Those who admit the possibility all agree that its realization would be very disagreeable. The same situation arises in connection with the possibility of our being superseded by some other animal species. This is almost as disagreeable and its theoretical possibility is indisputable.' Another reason he suggests is the 'religious belief that any attempt to construct such machines is a sort of Promethean irreverence ...we simply do not know of any creation which goes on creating itself in variety when the creator has withdrawn from it.'[40] Averting prejudice, Turing stated that 'we are not interested in the fact that the brain has the consistency of cold porridge. We don't want to say "This machine's quite hard, so it isn't a brain, and so it can't think."'[41] He concluded that 'the extent to which we regard something as behaving in an intelligent manner is determined as much by our own state of mind and training as by the properties of the object under consideration.'[42] Jackendoff takes a similar, non-Sapir-Whorfian approach:

> Our thoughts are revealed to us primarily through linguistic imagery, which is correlated with phonological structure. Of course, our visual and proprioceptive imagery also give us evidence of our thoughts, but my impression is that the average academic is overwhelmingly dominated by linguistic imagery, the unceasing voice in the head. (It might well be different for carpenters and painters and musicians, whose thinking takes place predominantly in domains that are better imaged nonlinguistically.)[43]

As Turing understood, we have the capacity to think in music, movement, colour, taste, abstractions, and so on. Einstein wrote in a letter to Jacques Hadamard that 'the words or the language, as they are written or spoken, do not seem to play any role in the mechanism of my thought.'[44] But based purely on introspection, one can see that much of our thinking involves language. Reviewing recent studies, David Robson concludes that 'up to 80 per cent of our mental experiences are verbal.'[45] The research of Jeffrey Loewenstein and Dedre Gentner suggests that 'the spatial reasoning of young children is improved by reminding them of words such as "top," "middle" and "bottom."'[46] Hearing a letter spoken out loud also helps an individual identify that letter amongst a cluster of others, suggesting a connection between language and perception.[47] Whether language evolved purely for the cognitive benefits is yet to be proven, but we can be sure that this original faculty of the mind helped our ancestors makes some sense out of the information their expanded brains were processing, allowing them to focus on the essential details of their surroundings.[48] Quine used to identify language as virtually the study of mind, but as we've demonstrated linguistic computation is only one aspect of our mental world (although a major one). The relation between language and the rest of cognition has been widely debated for centuries, but we do know, for instance, that our 'theory of mind' develops independently of language, since it's been found to be present in subjects who lost their language capacity.[49]

Despite the research of developmental psychologists and philosophers, the jury's still out (as it is in the case of music) on how close the connection between mathematics and language is, with both sharing many distinct qualities like recursion and unbounded Merge. The connection is certainly embedded in the mind, demonstrated by the variety of 'container' metaphors, sets and elements of metonymy both faculties possess. As the mathematician Richard Elwes puts it: 'We can deny the existence of infinity, a quantity that pervades modern mathematics, or we must resign ourselves to the idea that there are certain things about numbers we are destined never to know.'[50] For McGilchrist, the major neuroscientific evidence supports the claim that 'mathematical skills are divided between the hemispheres ... Addition and subtraction activate the right parietal lobe, whereas multiplication activates verbal remembrance of "times tables" in the left hemisphere. Calculating prodigies appear to use more right-hemisphere-dependent strategies, making use of episodic memory.'[51]

Though the ability of recursive (or discrete) infinity most likely comes from the language faculty, mathematical knowledge is not strictly dependent on language: 'Syntactic structure is distinct from logical structure: subjects that have lost their grasp of syntax following a left-hemisphere stroke remains able to use sophisticated thought processes, as complex as the structure of complex syntax, and can calculate and reason perfectly well.'[52] Neil Smith takes a broader, formalist perspective: 'Natural selection must have played a role in [the evolution of language], but so too have elementary physical constraints, such as the size of the human head. Many other factors have also, presumably, been in involved, such as the adoption of a trait in one domain for use in another: an example might be the exploitation of discrete infinity by both the number sense and the language faculty. Our knowledge is vanishingly small in this domain, the questions are coherent, the issues are empirical, and the problem of unifying the various bits of our knowledge with the rest of the natural sciences is standard.'[53] These factors have largely been ignored by biologists and linguists since the 'Cartesian linguists' in the seventeenth and eighteenth centuries; Freud's model of language, for instance, completely ignored syntax (even his principles of psychological and emotional development were mostly 'repressed,' quite different from 'unconscious'). The vastly ambiguous nature of language and the fact that we can translate (no matter how vaguely) from one language to another, are strong reasons to believe that there is an underlying property of the mind in which the different types of thought (visual, auditory, abstract, proprioception, etc.) interact. So 'perhaps the term "language" should be extended to other symbolic processes; perhaps introspection does not reveal the full extent to which the mind is speaking to itself.'[54] James McGilvray comments on the relationship between language and other mental 'modules':

> An account of *linguistic* concept acquisition need not explain acquisition of concepts in other faculties ... Our commonsense understanding of the world clearly depends on more than language. For instance, BUCKET draws shape and colour features from vision – facts that Marr emphasised when he pointed out that humans with damaged language centers retain a robust capacity to visually recognize buckets, although they cannot say what they are for. Language contributes (perhaps) CONTAINER, ARTIFACT, and USED TO MOVE MASSES OF NON-RIGID MATERIALS– the function of buckets, but not the abstract "shapes" Marr and others hold vision contributes.[55]

Neil Smith has also discussed the property of displacement (the ability to talk about things which are not immediately present), which 'may plausibly be motivated by the need to structure information for optimal communication. If this is, indeed, the correct account then it looks as if a property of the language faculty is imposed from outside the system, from another part of the mind-brain.'[56] Considering the intricacy of the language faculty, Pierre Jacob adds that it is possible that 'some piece of brain machinery (e.g., Broca's area) was selected for some function (e.g., the control of hand motions) and then later was recruited for some different function (i.e., the language faculty). Of course, the issue of whether a cognitive capacity is or not the direct result of natural selection is an entirely empirical issue, not a conceptual one.'[57] As we saw above, the faculty of language exists no more in isolation than the immune system does, interacting in highly complex ways with other 'mental organs.'

The Nobel laureate Murray Gell-Mann, having left physics to take up an earlier passion of his, linguistics, now spearheads the Evolution of Human Languages programme at the Santa Fe Institute. With possible relations to Chomsky and Pinker's research, Gell-Mann's programme has found 'tentative evidence for a situation in which a huge fraction of all human languages are descended from one spoken around 20,000 years ago, towards the end of the last ice age.'[58] Running parallel to this, geneticists have shown how there is a single 'master control gene' for all organic eyes in nature – superficially, all animal eyes are clearly different, but they have strong uniform ties beneath the surface.[59] This could well be the same with language, with the world's different varieties of it (or end states of I-languages) being superficially constructed over time from a universal internal structure:

> As for the "enormous variation" in structure of brains and experience, that tells us little. Not many years ago, languages appeared to differ from one another as radically as neural structures do to many a trained eye today, and were considered mere reflections of infinitely variable experience. Any complex system will appear to be a hopeless array of confusion before it comes to be understood, and its principles of organization and function discovered.[60]

Pinker thinks that 'probably in the evolution of the human species, evolution of language and the evolution of language in thought probably went together; each one helped the other. If you can think more complex thoughts, that puts pressure on you to be able to share them, and if you've got other people supplying you with complex language, that puts pressure on you to be able to have those thoughts.'[61] In the same way that the grammar of language emerges after sufficient stimulus from external sounds, so too are abstract mathematical representations elicited rather than 'learned.' For Carnap, 'asking whether there are really numbers is tantamount to asking whether the linguistic framework that governs number talk is "true" or "correct."'[62] Representationalist theories of mind, then, along with Kripke's 'rigid designator' and the virtually unshakable grip of the externalist religion of 'reference,' amount to forms of Pythagorean mysticism through which mathematics and language has a direct descriptive relation to the world.

Our common sense theory (or folk physics) of space, writes Colin McGinn, 'probably develops, in part, out of our perceptual systems and it serves to guide our behaviour; we might think of it as a visuo-motor space. No doubt it would be difficult to describe this mental representation of space in full detail, but I think it is fair to report that it encodes a broadly Euclidian geometry and that it regards motion as relative to the position of the earth. It also has some firm ideas about what it is for something to be somewhere.'[63] Since the time of Newton, and with Einstein's advances and the dawn of M-theory, our commonsense understanding of space and causation have been thrown into disarray: 'The status of causation,' John Collins points out, 'has been moot ever since Newton's impugnation of "hypotheses", notwithstanding the common appeal to the notion as if it were the natural relation par excellence.'[64] 'When Newton enunciated the law conception of cause,' wrote Tolstoy, 'mathematics seeks laws, that is, the property of attraction; he said that all bodies from the largest to the smallest have the property of attracting one another, that is, *leaving aside the question of the cause of the movement of the bodies*, he expressed the property common to all bodies from the infinitely large to the infinitely small.'[65] Zamyatin's *We* also explores similar vague intuitions surrounding infinity:

> "I understand you, I understand you very well,' he said. 'Nevertheless, you must calm yourself. Don't. All of this will return, it will inevitably return. The only important thing is that everyone must learn about my discovery. You are the first to hear it: according to my calculations, there is no infinity!"
> I stared at him wildly.
> "Yes, yes, I am telling you: there is no infinity. If the universe were infinite, then the mean density of matter in it should equal zero. And since it is not zero – we know that! – it means that the universe is finite; it is spherical in form, and the square of the cosmic radius, Y2, equals the mean density multiplied by ... Now this is the only thing I need – to compute the digital coefficient, and then ... You understand: everything is finite, everything is simple, everything is calculable. And then we shall conquer philosophically – do you understand? And you, my dear sir, are disturbing me, you are not letting me complete my calculation, you are screaming..."
> I don't know what shook me more – his discovery, or his firmness at that apocalyptic hour. In his hands (it was only now

that I noticed it) he had a notebook and a logarithmic table. And I realized that, even if everything should perish, it was my duty (to you, my unknown, beloved readers) to leave my notes in finished form.

I asked him for some paper – and it was there that these last lines were written...

I was about to put the final period to these notes, just as the ancients put crosses over the pits where they had thrown their dead, when suddenly the pencil shook and dropped from my fingers.

"Listen." I tugged at my neighbour. "Just listen to me! You must – you must give me an answer: out there, where your finite universe ends! What is out there, beyond it?"

He had no time to answer. From above, down the stairs – the clatter of feet...[66]

Rather than existing as Rorty's proclaimed 'mirror of nature,' our conceptualisation of certain mathematical ideas instead employ 'many cognitive mechanisms that are not specifically mathematical ... These include such ordinary cognitive mechanisms as those used for basic spatial relations, groupings, small quantities, motion, distributions of things in space, changes, bodily orientations, basic manipulations of objects (e.g., rotating and stretching), iterated actions, and so on.'[67] Though his Lakoffean 'conceptualist' argument for the nature of mathematics does not account for its actual emergence, only the cognitive mechanism used to employ it, Rafael Núñez nevertheless gives the following telling examples:

> Conceptualizing the technical mathematical concept of a class makes use of the everyday concept of a collection of objects in a bounded region of space.
> Conceptualizing the technical mathematical concept of recursion makes use of the everyday concept of a repeated action.
> Conceptualizing the technical mathematical concept of complex arithmetic makes use of the everyday concept of rotation.
> Conceptualizing derivates in calculus requires making use of such everyday concepts as notion, approaching a boundary, and so on.[68]

Writing on the philosophy of mathematics, Saunders MacLance comments that mathematics is 'not the study of intangible Platonic worlds, but of tangible formal systems which have arisen from real human activities.'[69] Aristotle's similar internalist perspective was that 'The general propositions in mathematics [the axioms and also the theorems of the Eudoxian theory of proportions] are not about separate things which exist outside of and alongside the [geometric] magnitudes and numbers, but are just about these; not, however, insofar as they are such as to have a magnitude or to be divisible.'[70] Stressing a related need for Newtonian simplicity, Heisenberg once pointed out to Einstein:

> If nature leads us to mathematical forms of great simplicity and beauty – by forms I am referring to coherent systems of hypothesis, axioms, etc. – to forms that no one has previously encountered, we cannot help thinking that they are 'true', that they reveal a genuine feature of nature ... You must have felt this too: the almost frightening simplicity and wholeness of the relationships which nature suddenly spreads out before us and for which none of us was in the least prepared.[71]

He also warned, regarding his brainchild quantum mechanics, 'we have at first no simple guide with correlating the mathematical symbols with concepts of ordinary language; and the only thing we know from the start is the fact that our common concepts cannot be applied to the structure of the atoms.'[72] Since the time of Schrödinger's thought experiment, which suggested that a cat could exist in a superposition of being both alive and dead, physicists have wondered whether large objects follow the laws of the quantum world.[73] Vlatko Vedral has written of some important developments in quantum theory that may shed some light on what part its counter-intuitive constituents play in the world at large. Although 'the world at large' is a misleading phrase: 'Somewhere between molecules and pears lies a boundary where the strangeness of quantum behaviour ends and the familiarity of classical physics begins. The impression that quantum mechanics is limited to the microworld permeates the public understanding of science.' This 'convenient partitioning of the world is a myth.' Quantum states are prone to collapse if their complexity reaches limits beyond the reaches of entanglement, but although 'quantum effects may be harder to see in the macroworld, the reason has nothing to do with the way that quantum systems interact with one another.' In fact these effects may function in the cells of our bodies; consequently the 'division between the quantum and classical worlds appears not to be fundamental.'[74] Considering the boundaries imposed on the natural sciences both by our obscure but existing cognitive limits and natural law, Kosso comments on the measurability (in principle) of a quark's 'flavour' but not its 'colour':

> To say that color cannot be seen is just a sloppy way to claim that a colored particle cannot interact with another object which we regard as part of the observing apparatus in such a way that the apparatus object is left in a different state according to

whether the interaction was with a red, green, or blue particle. The color of quarks is unobservable because physical interactions are colorblind. It is not a problem with our eyes or with our machines. The unobservability is dictated by the physical laws of interaction.[75]

The subject of time has also proved itself to defy intuition. When asked what his experience of time was like during a mescaline trip, Aldous Huxley replied: 'There seems to be plenty of it.'[76] String theorist Brian Greene, on other hand, might not agree with him:

> Time is with us, every moment. I can't even say a sentence without invoking a temporal word – moment. But what is time? When we look at the mathematics of what it is or where it came from, time is there, but there's no deep explanation of what it is or where it came from.[77]

Converge these views with our rejection of physicalism, and we find Galen Strawson writing in a recent essay on 'The Self': 'I should admit, though, that I don't fully know the nature of the physical. No one does. Nearly all of us take is that the physical is essentially spatio-temporal, for example, but no one expert in these matters claims to know for certain what space and time are, or whether they are really fundamental features of reality as we standardly conceive them.'[78] René Magritte, an artist with an eye for the illusory, held that 'We see [the world] as being outside ourselves, although it is only a mental representation of what we experience inside ourselves. ... Time and space thus lose that unrefined meaning which is the only one everyday experience takes into account.'[79]

For Dean Rickles, neither quantum physics nor general relativity can account for the existence of time: 'It is highly likely that what we think of as time emerges from some deeper, more primitive non-temporal structure.'[80] Linda Geddes adds that, 'While we have a fairly good grasp on the millisecond timing involved in fine motor tasks and the circadian rhythms of the 24-hour cycle, how we consciously perceive the passage of seconds and minutes – so-called internal timing – remains decidedly murky.' For humans 'there is no dedicated sensory organ for time perception, as there are for perceiving the physical and chemical nature of our environment through touch, taste and smell. Time is also unusual in that there is no clinical condition that can be defined purely as a lack of time perception, which makes it difficult to study.'[81]

Adding to this, Sean Carroll warns that 'We have no right to claim that the universe and time started at the big bang, or had some sort of prehistory.'[82] After the collapse of Cartesian mechanical philosophy, Leibniz, rivalling Newton, proposed that 'change is the fundamental property of the universe and that time emerges from our mental efforts to organise the changing world we see around us.'[83] Carlo Rovelli has also 'rewritten the rules of quantum mechanics so that they make no reference to time.'[84] Echoing Leibniz, he states that 'Physics is not about "how does the moon move through the sky?" but rather "how does the moon move in the sky with respect to the sun?" Time is in our mind, not in the basic physical reality.'[85]

But is there a similar Cartesian distinction between this 'physical reality' and Leibniz's 'mental efforts,' as is widely held both in philosophy and neurobiology? Let's take, as a starting point, the early formulation of this view: Descartes' influential mind-body separation, explained in a passage from his *Meditations*:

> I posses a body with which I am very intimately conjoined, yet because, on the one side, I have a clear and distinct idea of myself inasmuch as I am only a thinking and unextended thing, and as, on the other, I possess a distinct idea of body, inasmuch as it is only an extended and unthinking thing, it is certain that this I (that is to say, my soul by which I am what I am), is entirely and absolutely distinct from my body, and can exist without it.[86]

Do the mind and body therefore obey different physical laws? Hand pressed to forehead in despair over the question, Saul Kripke admits 'I regard the mind-body problem as wide open and extremely confusing.'[87] But attention, as always, needs to be payed to the terminology used to formulate this problem and the biology which rests beside it (what, after all, is meant by 'physical'?). Jaegwon Kim outlines the mystery as follows: 'There has been a virtual consensus, one that has held for years, that the world is essentially physical, at least in the following sense: if all matter were to be removed from the world, nothing would remain – no minds, no entelechies and no "vital forces."'[88] Princess Elizabeth of Bohemia, while in exile at The Hague, wrote to Descartes on June 20th 1643 rejecting his dualism, since 'it would be easier for me to attribute matter and extension to the soul, than to attribute to an immaterial thing the capacity to move and be moved by a body.'[89] Her response has been called 'the first causal argument for physicalism' by Kim.[90] Instead of holding onto any ontological commitments about what 'matter' is or is not capable of, Chomsky points out that 'we understand "the materialist world picture" to be whatever science constructs' – in its current mould it is, as Quine had it, 'theories of quarks and the like.' This should be understood no matter how drastically empirical evidence diverges from 'mechanical causes.' For questions cognitive or other, it would seem that full interpretation will be restrained until the constituent parts of mental or physical phenomena have been exposed for what they really are – something which, for reasons of our technological and biological limitations, currently eludes us:

> To put it differently, the discussions presuppose some antecedent understanding of what is physical or material, what are the physical entities. These terms had some sense within the mechanical philosophy [of Descartes]. But what do they mean in a world based on Newton's "mysterious force," or still more mysterious notions of fields of force curved space, infinite one-dimensional strings in ten-dimensional space or whatever science concocts tomorrow? Lacking a concept of "matter" or "body" or "the physical," we have no coherent way to formulate issues related to the "mind-body problems".[91]

Discussions of the mind-body duality are therefore ended with the realisation that the problem revolves not so much around the fact that we do not understand the concept of the 'mind,' but rather it is the concept of 'body' which lies in 'mystic obscurity,' to use Lange's term. It's curious to note that if the 'mental' is 'physical,' why is it that no one has posed a body-body problem? Today there no longer remains a coherent or fixed concept of 'body' or 'matter' (Einstein's general theory of relativity supposedly brought it back, but the later emergence of quantum physics finished it off for good): there remains only the world, 'with its various aspects: mechanical, electromagnetic, chemical, optical, organic, mental – categories that are not defined or delimited in an a priori way, but are at most conveniences: no one asks whether life falls within chemistry or biology, except for temporary convenience. In each of the shifting domains of constructive inquiry, one can try to develop intelligible explanatory theories, and to unify them, but no more than that.'[92] Science is suitably, for Stephen Gaukroger, 'a loose grouping of disciplines with different subject matters and different methods, tied in various ways of which work for some purposes but not for others.'[93] Geoffrey Hellman made similar observations in 1985:

> [C]urrent physics is surely incomplete (even in its ontology) as well as inaccurate (in its laws). This poses a dilemma: either physicalist principles are based on current physics, in which case there is every reason to think they are false; or else they are not, in which case it is, at best, difficult to interpret them, since they are based on a "physics" that does not exist – yet we lack

any general criterion on "physical object, property, or law" framed independently of existing physical theory.[94]

Priestley argued in favour of this proposal when he wrote that 'the whole man is of some *uniform composition*, and that the property of *perception*, as well as the other powers that are termed *mental*, is the result ... of such an organical structure as that of the brain.'[95] Neurons are, after all, in Gerald Feinberg's words, 'ordinary matter.' This is an explanation of the neurophysiological, and not a description of the mental. The very term 'mental,' in fact, should be understood in the same way the terms 'optical,' 'electrical,' and 'chemical' are – namely, categories within which certain aspects of the world are distinguished. There is no invisible line where physics begins and chemistry ends, rather each discipline is set by arbitrary boundaries to suit human interests and concerns.

Before he turned his attention to deploring the meddlesome 'swinish multitude' (or 'the public,' as they are nowadays referred to), Edmund Burke, like Oscar Wilde (for whom 'to define is to limit'), proposed in 1757:

> When we define we seem in danger of circumscribing nature within the bounds of our own notions, which we often ... form out of a limited and partial consideration of the object before us, instead of extending our ideas to take in all that nature comprehends, according to her manner of combining. ... A definition may be very exact, and yet go very little way towards informing us to the nature of the thing defined.[96]

Throughout the history of science it is commonly the more 'fundamental' science which has been forced to undergo revision for unification to take place. The unification of physics and chemistry is a modern example. 'Pauling's account of the chemical bond unified the disciplines, but only after the quantum revolution in physics made these steps possible. The unification of much of biology with chemistry a few years later might be regarded as genuine reduction, but that is not common, and has no particular epistemological or other significance; "expansion" of physics to incorporate what was known about valence, the Periodic table, chemical weights, and so on is no less valid a form of unification.'[97] A form of genuine reduction (or 'consilience,' to use G. O. Wilson's term) was the explanation of heat transfer through the statistics of molecular motion. Bohr, too, 'really had it in his mind that there was some profound problem with neutrinos and energy ... and did not want to have it solved except in a mystical and deep way. It was solved by Fermi in "too elementary" a way.'[98] This could also indicate that a radical alteration of our understanding of the neurophysical could be what's needed in order to unify our understanding of psychological states (like language) with neuroscience. Neuroscience has only recently emerged out of infancy, and it's yet to have its own Copernican-style revolution. But 'even if we know *where* a structure is localized in the brain – the sort of information that neural imaging can provide – we do not know *how* the brain initiates the structure.'[99] A neurolinguistic theory is 'incomplete if it does not offer genuine solution to the problems of combinatoriality, structural hierarchy, and binding among structures.'[100] More generally, 'talk of a lack of fit between physics and the theoretically intractable phenomena of consciousness and intentionality should strike us as odd, for it is not so clear if there are any uniform phenomena in view.' Additionally, 'the signature properties of the mental might well fragment under inquiry,' with 'the great variety of qualitative experience ... (compare the difference between hearing a noise and feeling a pain)' not appearing to be unified either phenomenologically or neurologically.[101] But the prospects of such interdisciplinary collaboration are bolstered by another simple observation of Jackendoff's, that 'of all the cognitive sciences, only linguistics has systematically and explicitly investigated the content of mental structures that underlie a human capacity.'[102] For Gilbert Harman, contemporary linguistics has is 'the most successful of the cognitive sciences.'[103] Lange reminds us too that 'We wonder at Newton's discovery of the law of gravitation, and scarcely reflect how much progress had to be made in order so far to pave the way for this doctrine that it must inevitably be discovered by some great thinker.'[104] Though we should be careful not to fall into the trap of intellectual complacency Democritus demonstrated, with his usual modesty:

> Among all my contemporaries, I have travelled over the largest portion of the earth in search of things the most remote, and have seen the most climates and countries, heard the largest number of thinkers, and no one has excelled me in geometric construction and demonstration – not even the geometers of the Egyptians, with whom I spent in all five years as a guest.[105]

In order for a complete account of the physical world, high-level principles and abstractions need to be accounted for by being 'grounded' in lower-level 'mechanisms,' Jeffrey Poland assures us, ignoring Chomsky's caveats. To quote Nobel Laureate Max Born on a similar impasse:

> How does it come about then, that great scientists such as Einstein, Schrödinger and De Broglie are nevertheless

dissatisfied with the situation? Of course, all these objections are levelled not against the correctness of the formulae, but against their interpretation. ... The lesson to be learned from what I have told of the origins of quantum mechanics is that probable refinements of mathematical methods will not suffice to produce a satisfactory theory, but that somewhere in our doctrine is hidden a concept, unjustified by experience, which we must eliminate to open up the road.[106]

Mental properties may not, therefore, be able to be reduced to what Patricia Churchland calls 'neural-network properties.'[107] Newton taught us that bodies are not static, inert entities floating in space, and so we are quite within reason to conclude that matter 'is no more incompatible with sensation and thought than with attraction and repulsion.'[108] Ideas themselves – which Locke defined as powers to produce certain sensations 'in us' (one of his favourite, ambiguous, phrases) – are not theoretical entities, but common-sense, 'folk scientific' ones. Locke's central psychological argument (not 'epistemological,' as is commonly claimed, with Locke stressing the close relation between his theories and physiology) has also been identified by Hinzen to be a 'paradigmatically non-empiricist contention,' contrary to illusion. He argued in the *Essay* that there is no '*necessary* connection between "the bulk, figure, and motion of several Bodies about us" and the sensations or ideas these produce in us,' a plainly internalist (though not quite rationalist) assumption about mental structure.[109]

Agreeing with Poland, John Searle, among many respectable biologists, believes consciousness is 'a higher-level or emergent property of the brain.' It is 'as much of the natural biological order as ... photosynthesis, digestion, or mitosis.'[110] In the early 1970s, Philip Anderson championed the view of 'emergence,' or 'the notion that important kinds of organisation might emerge in systems of many interacting parts, but not follow in any way from the properties of those parts.'[111] But as Chomsky cautions: 'There's a slogan – that the mind is neurophysiology at a more abstract level. But chemistry wasn't physics at a more abstract level [in the early 1920s], as it turned out. Rather, a new physics came along, which was chemistry at a different level. And we don't know that that won't happen in the study of mind.'[112] The related error of type-identity physicalism, according to Andrew Melnyk, again revolves around its claim that 'every kind of thing spoken of in any science *is identical* with some physical kind of thing.'[113]

Chris Daly was also ahead of most contemporary philosophers when he pointed out in 1998 that by lacking a concept of physical 'no debate between physicalism and dualism can even be set up.'[114] The post-Newtonian world simply doesn't entertain such 'material' notions. The conceptual implications are laid out by Pinker, who reminds us that while quantum physics is infamously counterintuitive,

> What is less appreciated is that classical Newtonian physics is *also* deeply counterintuitive. The theory in the history of physics that is closest to intuitive force dynamics is the medieval notion of impetus, in which a moving object has been imbued with some kind of vim or zest that pushes it along for a while and gradually dissipates.[115]

There is consequently a vast gulf between folk science and 'real physics, which is just a bunch of differential equations specifying how objects change their velocity over time.'[116] Folk science has been viewed as 'reality' by many since the origins of science, as Lange points out:

> How content were our forefathers on their earth, resting in the bounded sphere of the eternally-revolving vault of heaven, and what agitation was excited by the keen current of air that burst in from infinity when Copernicus rent this curtain asunder![117]

The folk scientific and linguistic notion of agency is also integral to 'causation,' as Carl Sagan demonstrated in his intentional description of the Big Bang: 'If you want to make an apple pie from scratch, you must first create the universe.' But, as Russell tells us, the First Cause argument 'does not carry very much weight nowadays, because, in the first place, cause is not quite what it used to be.'[118] Lange wrote of Epicurus in his *History of Materialism* that 'In his fourteenth year, it is said, he studies Hesiod's *Cosmogony* at school, and finding that everything was explained to arise from chaos, he cried out and asked, Whence, then, came chaos? To this his teacher had no reply that would content him, and from that hour the young Epikurus began to philosophise for himself.'[119] He later concluded, to the chagrin of later theologians, that 'even the motion of the heavenly bodies is not dependent upon the wish or impulse of a divine being; nor are the heavenly bodies themselves divine beings, but everything is governed by an eternal order which regulates the interchange of origination and destruction.'[120] In his *Mysticism and Logic*, Russell starts a similar discussion, claiming that 'the reason why physics has ceased to look for causes is that in fact there are no such thing. The Law of Causality, I believe, like much that passes muster among philosophers, is a relic of a bygone age, surviving, like the monarchy, only because it is erroneously supposed to do no harm.'[121] Russell's colleague, Alfred North Whitehead, had also written decades earlier that 'There is no such thing as nature at an instant posited by sense-awareness. ... What we perceive as present is the vivid fringe of memory tinged with anticipation.'[122]

During his description of the cosmological constant in his entry in the *Dictionary of Philosophy*, Anthony Flew goes beyond Epicurus and observes that any attempt 'to explain why things are as they are must always ultimately be made in terms of general facts that are not, and cannot be, further explained. So why should the existence of the Universe, and perhaps the fact that it has whatever fundamental regularities it does have, not be accepted as the fundamentals, requiring no further explanation.'[123] Besides which, there are simple virtues to the physicist's view of the world which all others lack, as Lange again notes:

> There are peoples which believe that the earth rests upon a tortoise, but you must not ask on what the tortoise rests. So easily is mankind for many generations contended with a solution which no one could find really satisfactory. By the side of such fantasies the creation of the world from nothing is at least a clear and honest theory. It contains so open and direct a contradiction of all thought, that all weaker and more reserved contradictions must feel ashamed beside it.[124]

Since we 'attribute "reality" to whatever is postulated in the best theory we can devise' (Chomsky), and since the post-Newtonian world lacks a fixed concept of 'physical' or 'matter,' questions of 'reality' are reduced simply to questions of theories. 'Matter' and 'physical' are simply terms which stand for whatever entities are postulated by present naturalism, with any Cartesian intuitions about mechanical causes failing to explain the possible existence of Majorana particles, which are simultaneously matter and anti-matter (even the fact that matter is mostly empty space does not deny its place amongst 'the physical': 'Anti-matter' is still 'matter,' so to speak.).[125] John Ziman's influential *Public Knowledge* recognises that 'The objective of science is not just to acquire nor to utter all non-contradictory notions: its goal is a consensus of rational opinion over the widest possible field' – with 'consensus' being the term metaphysically closest to 'reality.'[126]

Understanding the need to construct mathematical theories which defy our intuitions (as the linguists and physicists do), Galileo praised those who 'did violence to their senses' in his *Dialogo*.[127] Seeing the rationalist tradition of the seventeenth and eighteenth centuries as precursors to the minimalist program, Hinzen notes that, 'As Galileo had assumed, ordinary things have a complexity that escapes mathematisation. To understand anything, one has to look away from the world of the senses and employ *idealizations* that *distort* nature as known to us.'[128] There are, furthermore, certain levels of complexity which preclude theory-formation. We cannot, for example, abstract from experience the Battle of Waterloo, observe and collect data (number of bullets fired, average height of soldier, etc.) and expect the data to tell us even a fraction of what actually occurred; nor could we assume that our data would allow us to accurately predict the nature of future warfare (though it would certainly give a more accurate prediction than the physicists). General correlations may well occur, but the strict disciplinary methods which naturalistic inquiry observes and the practice of conducting closed experiments would be impossible for something involving as many factors as a military engagement. Derek Fraser's study *The Evolution of the British Welfare State* is consequently not the kind of evolution Darwin could have studied.[129] Galileo's beliefs were reformulated by Weinberg in 1976, who spoke of 'abstract mathematical models of the universe to which at least the physicists give a higher degree of reality than they accord the ordinary world of sensation.'[130] Leibniz, on the other hand, denounced Newton's gravitational force as 'an occult quality' so incomprehensible and 'so very occult, that it is impossible that it should ever become clear.'[131] Contemporary physicists may not agree with Leibniz's conclusion, but his premise of incomprehensibility has become standard doctrine in the physical sciences ever since.

Since post-Newtonian science strives for intelligibility of theories and not intelligibility of the world (as Chomsky has recently put it), and since 'immaterial' forces and not purely mechanical causes are the object of inquiry, science (or 'methodological naturalism,' for Hinzen) 'is just "natural philosophy" in the original sense of Galileo, Locke, Hume, or Descartes.'[132] Correspondingly, ontological questions about what things 'really are' are left aside as a hindrance, of no more concern to cognitive science than the classical categories of 'earth', 'air', 'fire' and 'water' are to physics. We are in a similar position as the medieval scholastics, with 'the world' once again beyond comprehension, a feeling Plato touched on when musing on the accumulation of knowledge: 'Every one of us is like a man who sees things in a dream and thinks that he knows them perfectly and then wakes up, as it were, to find that he knows nothing.'[133]

But in many crucial respects there has been considerable regression in the sciences of the mind since the early nineteenth century, spurred on by the division between 'science' and 'philosophy.' Only after Kant was philosophy independently institutionalized as an academic discipline. Though this has helped organise inquiry into the many domains previously termed 'natural philosophy,' Huxley's thoughts reminds us of the damage caused by too much compartmentalisation:

> Literary or scientific, liberal or specialist, all our education is predominantly verbal and therefore fails to accomplish what it is supposed to do. Instead of transforming children into fully developed adults, it turns out students of the natural sciences who are completely unaware of Nature as the primary fact of experience, it inflicts upon the world students of the humanities who know nothing of humanity, their own or anyone else's.[134]

The seventeenth and eighteenth century theories of perception could be seen, then, as the 'philosophy of vision' (perhaps the fact that the words 'philosophy of' are predominantly attached only to studies of language and mind reveals the methodological dualism still gripping each domain). '"Naturalism" in philosophy was no issue; philosophy as such was methodologically naturalistic, and no argument for any such naturalism was asked for.'[135] This is something Joseph Priestley understood well: 'In all investigations relating to human nature, the philosopher will apply the same rules by which his inquiries have been conducted upon all other subjects. He will attentively consider appearances, and will not have recourse to more causes than are necessary to account for them.'[136] It's for this reason that Carnap and Quine's advocation of reducing philosophy to science through introducing the 'philosophy of science' has been entirely misleading and damaging (not that there aren't enough fine philosophers of science making important contributions to the physical sciences), since it presupposes two separate methodologies and areas of inquiry: the 'scientific method' and, even more obscure, the 'philosophical method.' Many philosophers like Quine, Carnap, Toulmin and Popper first took degrees in physics or mathematics, only later turning to philosophical questions, their love of a priori reflection presumably intact. Terms like 'philosophy,' 'science,' 'matter' and 'mind' are largely historical residues, and considering them in isolation by ignoring their origins only leads to confusion. 'Armchair thinking is a wonderful thing,' argue the authors of a standard textbook of cognitive neuroscience, 'and has produced fascinating science such as theoretical physics and mathematics.'[137] But do we therefore conclude that Stephen Hawking is a philosopher? Surely not, with J. J. Thomson also spending 'a good part of most days in the armchair of Maxwell doing mathematics.'[138] Paul Dirac also believed most of his time was spent 'just playing with equations and seeing what they give.'[139] Derek Gjersten reminds us of 'the obvious but frequently ignored point that much of science involves no more than thinking,' and that science can even be studied as 'a system of informed thought,' though often going beyond pure reflection.[140] The 'wonderful' nature of armchair reflection was understood by Cato, who according to Cicero used to say that 'never is a man more active than when he does nothing, never is he less alone than when he is by himself.'[141] According to Marjorie McCorquodale, 'studies by Rollo May, Bronowsky, and others indicate that while actually engaging the creative process of scientific discovery or in that of imaginative writing, the minds of scientists and poets work in remarkably similar, almost identical ways.'[142] And so the 'sciences of life can confirm the intuitions of the artist, can deepen his insights and extend the range of his vision.'[143] There really is no defined, stable scientific 'method' other than simply being reasonable. A psychology degree, for example, includes a 'methodology' course, but a physics degree doesn't. There are, of course, different kinds of theories within the sciences, some stressing the importance of natural selection (evolutionary psychology), others on physical constraints (theoretical biology), others on 'elegance' and 'beauty' (theoretical physics, cosmology), but none of them cohere mysteriously by following the same 'method' any more than artists follow the same 'aesthetic method' when painting portraits or crafting sculptures.

'Philosophy' can therefore be regarded as 'thoughtful' or 'reflective science,' or the subject of those aspects of the world open more to a priori exploration than experiment; say, ethics. It often depends on what people want to call themselves: note the virtual uniformity in subject matter between the work of Dennett, Harris, Pinker and Dawkins, yet they range from, at various points of their careers, from calling themselves philosophers, neuroscientists, linguists, experimental/evolutionary psychologists and biologists. John D. Barrow, the English cosmologist and mathematician, has written like Huxley that 'One can take two views of the activities of scientists like physicists who are seeking things they call the fundamental laws of nature. Either you believe, as they often do, that they are discovering the whole thing, and that one day we will hit on the mathematical form of the ultimate laws of nature. Alternatively one may be more modest and regard the scientific enterprise as an editorial process in which we are constantly refining and updating our picture of reality using images and approximations that seem best fitted to the process.'[144] McGilchrist has also noted how the boundaries of understanding have often been pushed in unexpected ways:

> In fact we know, though scientific method plays its part, the greatest advances of science are often the result of chance observations, the obsessions of particular personalities, and

intuitions that can be positively inhibited by too rigid a structure, method or world view. Technical advances, too, have been less often the foreseen consequences of systematic method than the results of local enthusiasts or skilled artisans attempting empirically to solve a local problem, and many have been frankly serendipitous by-products of an attempt to achieve something quite different.[145]

It's of little surprise, then, that Poincaré developed some of his most important mathematical ideas while on holiday at the seaside.[146] Doubtless approving of such behaviour, E. O. Wilson expressed his belief that

The best of science doesn't consist of mathematical models and experiments, as textbooks make it seem. Those come later. It springs fresh from a more primitive mode of thought, wherein the hunter's mind weaves ideas from old fact and fresh metaphors and the scrambled crazy images of things recently seen. To move forward is to concoct new patterns of thought, which in turn dictate the design of the models and experiments.[147]

Elaborating on a similar anarchist conception of scientific inquiry, Humboldt wrote in 1792 that 'all moral cultures spring solely and immediately from the inner life of the soul and can never be produced by external and artificial contrivances. The cultivation of the understanding, as of any of man's other faculties, is generally achieved by his own activity, his own ingenuity, or his own methods of using the discoveries of others.'[148] Radical changes have continuously proven to be the work of an individual or group thinking outside the usually accepted structure of inquiry. Newton, for one, was shocked and perplexed by his findings; something Huxley would have responded to by writing optimistically: 'But sooner or later the necessary means will be discovered, the appropriate weapons will be forged, the long-awaited pioneer of genius will turn up and, quite casually, as though it were the most natural thing in the world, point out the way.'[149]

Like Gjersten, Otto Neurath reminds us that, contrary to popular wisdom, 'There is no scientific method. There are only scientific methods. And each of these is fragile; replaceable, indeed destined for replacement; contested from decade to decade, from discipline to discipline, even from lab to lab.'[150] Sir Humphry Davy famously emphasised the importance of such ingenuity: 'Nothing is so fatal to the progress of the human mind as to suppose that our views of science are ultimate; that there are no mysteries in nature; that our triumphs are complete, and that there are no new worlds to conquer.'[151] The advance of science in twentieth century was so dramatic and sweeping that Lord Kelvin's words in 1900 should force us to take stock of the vast number of groundbreaking discoveries which still await human curiosity: 'There is nothing new to be discovered in physics now. All that remains is more and more precise measurements.' Auguste Comte, the philosopher and founder of sociology, made the comparable claim in 1835 that 'We shall never be able to study, by any method' the 'chemical composition' of the stars, nor 'their mineralogical structure.'[152] And, at the early stages of quantum physics Arthur Eddington proclaimed, 'We know nothing about the intrinsic nature of space.'[153]

If one were to ask Newton whether he was a philosopher or a scientist, he wouldn't have been able to answer. The term 'natural philosophy' was applied to Hume, Berkeley, Locke and Newton. 'In those dark times,' writes John Aubrey, 'astrologer, mathematician and conjurer were accounted the same things.'[154] But the science of the nineteenth century proved to be too complicated and obscure for all but a few specialists, and so 'scientists' diverged from 'physical' and 'moral' philosophers. In the 1970s, Joseph Campbell gave the following justification for studying religion and figurative language:

> When these stories are interpreted, though, not as reports of historic fact, but as merely imagined episodes projected onto history, and when they are recognized, then, as analogous to like projections produced elsewhere, in China, India, and Yucatan, the import becomes obvious; namely, that although false and to be rejected as accounts of physical history, such universally cherished figures of the mythic imagination must represent facts of the mind.[155]

In many ways, then, religion too was early science. It was our first assessment of the universe and our place in it, our first attempt at cosmology, our first attempt at a healthcare service, our first attempt at moral philosophy, among many other things. Or, as Aldous Huxley put it, 'Before the rise of science, the only answers to these questions came from the philosopher-poets and the poet-philosophers.'[156] Religion is also, as Strawson puts it, 'one of the fundamental vehicles of human narcissism.'[157] 'What are the stars', asked Thomas Pynchon, 'but points in the body of God where we insert the healing needles of our terror and longing?'[158] The reason our first attempts at philosophy and cosmology (in the form of religious texts and practices) have survived for so long is mainly due to the fact that they were precisely that – our first attempts. Friedrich Lange spots a similar intuition in Lucretius' magnificent *On the Nature of the Universe*:

> Religion is traced by Lucretius to sources that were originally pure. Waking, and still more in dreams, men beheld in spirit the noble and mighty figures of the gods, and assigned to these pictures of fancy, life, sensation, and superhuman powers. But, at the same time, they observed the regular change of the seasons, and the risings and settings of the stars. Since they did not know the reason of these things, they transferred the gods into the sky, the abode of light, and ascribed to them, along with all the celestial phenomena, storm also and hail, the lightning flash, the growling, threatening thunder.[159]

Taking the 'theory of mind' into account, Pinker continues that 'Perhaps the ubiquitous belief in spirits, souls, gods, angels, and so on, consists of our intuitive psychology running amok. If you are prone to attributing an invisible entity called "the mind" to other people's bodies, it's a short step to imagining minds that exist independently of bodies. After all, it's not as if you could reach out and touch someone else's mind; you are always making an inferential leap. It's just one extra inferential step to say that a mind is not invariable housed in a body.'[160] Following in the footsteps of Newton's 'occult force,' Hume believed we can know nothing of the ultimate nature of 'Causation': 'The power of force, which activates the whole machine ... of the universe ... is entirely concealed from us.'[161] To the intuitive objection that the tides could be controlled by the moon, Hume responded:

> Though distant objects may sometimes seem productive of each other, they are commonly found upon examination to be linked by a chain of causes, which are contiguous among themselves, and to distant objects; and when in any particular instance we cannot discover this connection, we still presume it to exist.[162]

Though such 'occult forces' are now a specialty of the physical sciences, Newton was initially criticised by Leibniz, Huygens, and Berkeley for reintroducing 'inexplicable qualities' into nature, which Descartes' mechanical philosophy had erased. Materialism, as Russell pointed out in his introduction of Lange's classic *History of Materialism*, has enjoyed a 'curious history.'[163] Many philosophers hang on to a physical notion of causality, mistaking their folk science for metaphysics, not realising that any such notion disappeared along with the mechanical philosophy, as Hume observed. David Papineau, for one, maintains the classical notion of causation in a recent essay on naturalism:

At first pass the causal closure of physics says that every physical effect has a sufficient physical cause. If this thesis is true, it distinguishes physics from all other subject domains. The biological realm, for example, is not causally closed in this sense, since biological effects often have non-biological causes, as when the impact of a meteorite precipitated the extinction of the dinosaurs.[164]

Like Papineau, even 6-month old infants experience illusions of causality, with older infants expecting causal interactions between inanimate objects since they have an innate 'principle of contact,' in Spelke's terms (the probable origin of Descartes' contact mechanics, demolished by Newton).[165] Adults (and, it's now known, 3-4 year olds[166]) also attribute intentional descriptions such as chased, attacked or comforted to moving geometrical shapes (the angry square chased after the little triangle, etc.).[167] Parmenides would have most likely agreed with our materialist assumptions so far, pre-dating Descartes' stance on subjectivity by concluding: 'Therefore thinking and that by reason of which thought exists are one and the same thing, for thou wilt not find thinking without the being from which it receives its name.'[168] Stephen Stich explains that with 'the decline of Cartesian dualism, philosophers began looking for a way to locate the mental *within* the physical, identifying mental events with some category of events in the physical world.'[169]

Hilary Putnam once pointed out that 'even the simple fact that a square peg won't fit into a round hole cannot be explained in terms of molecules and atoms but only at a higher level of analysis involving rigidity (regardless of what makes the peg rigid) and geometry.'[170] In addition to Chomsky's point, Gell-Mann has said that reduction in the natural sciences as a method of unification is 'great, but it will only take you so far in the study of complex subjects. Do you try to understand earthquakes in terms of quarks? Of course not. You use intermediate concepts, like plate tectonics and friction.'[171] This is, in many ways, an answer to Russell's question in his essay 'Why I Took to Philosophy,' when he asked his reader 'are things really what they look like to the naked eye or what they look like through a microscope?'[172]

Priestley, as one of the most influential chemists of his age, thought the Cartesian theory that reasons that a man is 'capable of thinking better when the body and brain are destroyed, seems to be the most unphilosophical and absurd of all conclusions.'[173] Keats, Shelley and Coleridge, along with other Romantic poets, were heavily influenced by this naturalistic account of the mind. Priestley later came to the accurate conclusion that there exists no natural law which makes matter incompatible with consciousness any more than there is one which makes it incompatible with attraction and repulsion. Priestley's position was that 'the powers of sensation or perception and thought, as belonging to man, have never been found but in conjunction with a certain organized system of matter; and therefore, that those powers necessarily exist in, and depend upon, such a system. This, as least, much be our conclusion, till it can be shown that these powers are incompatible with other known properties of the same substance; and for this I see no sort of pretence ... we ought to conclude that the whole man is material unless it should appear that he has some powers or properties that are absolutely incompatible with matter.'[174] Though accepting the possible need for a new understanding of the neurophysical, David Lightfoot, in a recent and important essay on 'Natural Selection-Itis,' perhaps falls prey to the metaphysical delusions over 'the mental' Priestley sought to combat: The evolution of language, he argues, is 'the toughest problem in all of science,' not least because 'we do not understand how brain matter secretes consciousness and communication and we need the help of people who study matter; in fact, we need a new understanding of "physiology", where the notion incorporates mental aspects.'[175] Nodding to the formalists, he also considers Dobzhansky's maxim that nothing in biology makes sense except in the light of evolution:

> It may be that current attempts to formulate "biolinguistic" analyses cannot yet make sense in the light of evolution, because we do not know enough to specify either the physical constraints on brain genomes that might have allowed the current genome to evolve from its predecessor, nor how that innovation might yield the new behaviour, nor the circumstances under which the evolutionary development took place.[176]

Priestley also demolished Newton's posited 'aether' by reducing matter to 'powers,' 'concluding that "solid matter" was an illusion and there was no "substration" of matter independent of its "powers". The distinction between matter and spirit has collapsed.'[177] 'Man,' thought Marx Wartofsky, 'is certainly a physical entity, and as a body moving in space or as a heat-engine his structure and processes are describable in terms of mechanics and thermodynamics.'[178] Adding to his believe that there exists inconceivable effects in motion, Locke (forming what is now known as 'Locke's suggestion') wrote that it is 'not much more remote from our comprehension to conceive that GOD can, if he pleases, superadd to matter a faculty of thinking, than that he should superadd to it another substance with a faculty of thinking.'[179] Like

Priestley, Dewey wrote how, 'given that consciousness exists at all, there is no mystery in its being connected with what it is connected with.'[180] Or, as Darwin put it, 'Why is thought, being a secretion of the brain, more wonderful than gravity, a property of matter?'[181]

Francesco Algarotti, translating Newtonian physics for the unversed (specifically 'the Ladies,' in his case), also acknowledged in 1737 that 'we are as yet but Children in this vast Universe, and are very far from having a compleat Idea of Matter; we are utterly unable to pronounce what Properties are agreeable to it, and what are not.'[182] Concise and insightful as always, Galen Strawson notes that we know nothing about the physical which should lead us to doubt that experiential phenomena are wholly physical phenomena: 'You might as well think that the efficacy of the binary system raises doubts about the validity of the decimal system.'[183] Ever since Boltzmann's time the notion of certainty hasn't existed in the physical sciences, with probability reigning supreme. What hasn't been acknowledged by either physicists or philosophers, however, is the possibility that there are aspects of the universe which fall outside of determinacy and probability, like, for instance, the Cartesian 'creative' use of language which, as Chomsky often stresses, is appropriate to situations but isn't caused by them. Language use, then, may not be a fit topic for naturalistic inquiry.

Drawing on Lewis Carroll's unmatched 'indefiniteness' in his poem 'Jabberwocky,' Eddington laid out the following thoughts on the achievements of the physical sciences:

> *Something unknown is doing we don't know what* – that is what our theory amounts to. It does not sound a particularly illuminating theory. I have read something like it elsewhere...
> ...the slithy toves
> Did gyre and gimble in the wabe.
> There is the same suggestion of activity. There is the same indefiniteness as to the nature of the activity and of what it is that it is doing. And yet from so unpromising a beginning we really do get somewhere. We bring into order a host of apparently unrelated phenomena; we make predictions, and our predictions come off. The reason – the sole reason – for this progress is that our description is not limited to unknown agents, executing unknown activities, but *numbers* are scattered freely in the description. To contemplate electrons circulating in the atom carries us no further; but by contemplating eight circulating electrons in one atom and seven circulating atoms in another we begin to realise the differences between oxygen and nitrogen. Eight slithy toves gyre and gimble in the oxygen wabe; seven in nitrogen.[184]

In Newton's time, to take another example, electricity was seen as a mystical and transcendent force, entirely irreducible to matter. With a deeper understanding of the neurophysical, then, perhaps the more mysterious nature of consciousness will be unified, like electromagnetism, with the natural sciences. As Priestley would likely confirm, a mind-body distinction is, certainly for today's science, rather like a wave-water distinction. Further insight into the physical (and, by extension, the neurophysiological) could instead be what is needed to explain mental faculties, and not a re-working of our understanding of language, memory, intentionality and so on.

The process of unification and reduction are often reflexive in their extension of knowledge – meaning, by revealing more about complex mental structures, like grammar and memory, we can invariably shed light on what the more fundamental components are up to. McGilchrist echoes this perspective, since 'even if it were possible for mind to be "reduced", as we say, to matter, this would necessarily and equally compel us to sophisticate our idea of what matter is, and is capable of becoming, namely something as extraordinary as mind.'[185] To put it another way, seeking a grand unified theory of the universe is 'the problem of constructing a system of theories and entities with exhibits the idealized structure characterized by physicalist theses.' Local unification problems between two smaller-scale disciplines (like the convergence of chemistry and physics in the early twentieth century) 'concern specific ontological or explanatory relations between more or less adjacent theories and entities within the larger system,' in the words of Jeffrey Poland. The advent of a Grand Unified Theory will only be arrived at after the unification of multiple 'local unification problems.'[186] Consequently, as Wartofsky concludes, 'The persistence of distinctive biochemical, biophysical, or biological formulations would only be a practical expedient, marking the difficulty to accomplishing the full reduction to physics, because of the great complexity of structure of living things' – bringing with it the possibility that certain biological laws may not be reducible to mechanistic interpretation.[187]

Problems of unification aside, Hume's speculation was that 'The mind is a kind of theatre, where several perceptions successfully made their appearance; pass, repass, glide away, and mingle in an infinite variety of postures and situations.' This is initially consistent with the orthodox mind-body division; however he added that this metaphor is not quite accurate, for the simple reason that there is no stage: 'They are the successive perceptions only, that constitute the mind; nor have we the most distant notion of the place, where these scenes are represented, or of the materials, of which it is compos'd.'[188] This is conceivably a reversal of what Russell was told as a child, namely that philosophy could be summed up in a single sentence: 'What is mind? – No matter. What is matter? – Never mind.'[189] In the current climate of computational approaches to cognition, the mind is viewed not as a Cartesian theatre but is 'algebraic' instead, in Marcus' sense.[190]

It's likely that part of the reason why the mind-body distinction has caused such problems revolves around our adapted proficiency at certain everyday tasks, so much so that we perform such acts with unconscious precision, with a feeling of detachment arising from our body's movements. But as Lawrence tells us:

> Why should I look at my hand, as it so cleverly writes these words, and decide that it is a mere nothing compared to the mind that directs it? Is there really any huge difference between my hand and my brain? – or my mind? My hand is alive, it flickers with a life of its own. It meets all the strange universe, in touch, and learns a vast number of things, and knows a vast number of things. My hand, as it writes these words, slips gaily along, jumps like a grasshopper to dot an i, feels the table rather cold, gets a little bored if I write too long, has its own rudiments of thought, and is just as much me as is my brain, my mind or my soul. Why should I imagine that there is a me which is more me than my hand is?[191]

Einstein also seems to have rejected this conceptual distinction between self-awareness and self: 'We may therefore regard matter as being constituted by the regions of space in which the field is extremely intense ... There is no place in this new kind of physics both for the field and matter, for the field is the only reality.'[192] Dirac's equation for the quantum-mechanical electron has also been approached by Giovanni Vignale in the following way: 'Dirac had started from a single particle, but in the attempt to pin it down to one point he found that he had to generate many more particles (some of them actually antiparticles). The inevitable conclusion was that a single particle does not exist in the literal sense, but only as a metaphor to describe a far more complex state of infinitely many particles and antiparticles – that is, a quantum field.'[193] Lowe, Richard C. Allen and Chomsky give three further insightful explanations:

> Newtonian inertial mass is definable in terms of a body's disposition to accelerate under the action of a force, in accordance with Newton's Second Law of Motion: the less a body accelerates under the action of a given force, the more massive it is (and this provides us with a measure of the *magnitude* of its mass). But the trouble is that the applicability

> of any such definition presupposes that we already have an adequate conception of what constitutes a *body*, and this is precisely what is now at issue. If all we know about 'bodies' is that they are *massive* objects with shape, size and motion, it helps us not at all to be told that mass is a dispositional property which a *body* has to accelerate at a given rate under the action of a given force.[194]

> In the Newtonian model, the supersaturated solution that is the Cartesian ocean of matter collapses, precipitating out crystalline lattices that contain infinitesimal quantities of matter and leaving behind the vast emptiness of absolute space.[195]

> [T]he problem of explaining terrestrial and planetary motion in terms of the 'mechanical' philosophy and its contact mechanics, demonstrated to be irresolvable by Newton, and overcome by introducing what were understood to be 'immaterial forces'; the problem of reducing electricity and magnetism to mechanics, unsolvable and overcome by the even stronger assumption that fields are really physical things; the problem of reducing chemistry to the world of hard particles in motion, energy, and electromagnetic waves, only overcome with the introduction of even weirder hypotheses about the nature of the physical world.[196]

The process has a taken a long time, and Aldous Huxley's words spread sensible caution, not of science's nature, but of what its eyes are constantly fixed on: 'For Science in its totality, the ultimate goal is the creation of a monistic system in which ... the world's enormous multiplicity is reduced to something like unity, and the endless succession of unique events of a great many different kinds get tidied and simplified into a single rational order. Whether this goal will ever be reached remains to be seen.'[197] Nature, as Heraclitus tells us, 'loves to hide.' Margaret Jacob writes that as a result of the scientific revolution of the seventeenth century, later physics was grounded in the view 'based on one reading of Newtonian physics, that motion is inherent in matter, that all of nature is alive, that soul and body are one, all material, all entirely of this world.'[198] As folk science dictates, the doctrine of the Cartesian mechanical philosophy proposed that 'each thing, in so far as it is simple and undivided, always remains in the same state as far as it can, and never changes except as a result of external cause.'[199] Along with theories like Kepler's, it was believed by most in the seventeenth century that the universe was built like a machine, with parts lacking any intrinsic relationship to each other. Having recognised this, contemporary 'natural philosophers' can no longer cling to such intuitive contact mechanics:

> The mind-body problem can be posed sensibly only insofar as we have a definite conception of body. If we have no such definite and fixed conception, we cannot ask whether some phenomena fall beyond its range. The Cartesians offered a fairly definite conception of body in terms of their contact mechanics, which in many respects reflects commonsense understanding. Therefore they could sensibly formulate the mind-body problem.[200]
>
> Matter and mind are not two categories of things, but they may pose entirely different kinds of quandaries for human intelligence, a fact that is interesting and important, if true, but in no way surprising to the naturalistic temper, which takes for granted that humans will face problems and mysteries, as determined by their special nature.[201]

For Galen Strawson, 'physical' is 'the ultimate natural-kind term,' and as the natural science's body of doctrine the notion increases, so the notion 'physical' expands. But natural kinds themselves are simply different types of concepts, differing from what naturalistic inquiry tells us about the world. Differences also occur even between what a geologist tells us compared to what a physicist tells us. For Michael Friedman,

> the philosophers of the modern tradition are not best understood as attempting to stand outside the new science so as to show, from some mysterious point outside of science itself, that our scientific knowledge somehow "mirrors" an independently existing reality. Rather, [they] start from the *fact* of modern scientific knowledge as a fixed point, as it were. Their problem is not so much to justify this knowledge from some "higher" standpoint as to articulate the new *philosophical* conceptions that are forced upon us by the new science.[202]

The very existence of the term 'physical' lays the foundation for one to exercise the human tendency for negation – namely, to invent the concept of 'anti-physical' or 'un-physical,' provoking and increasing the likelihood of belief in 'non-physical' entities like 'God' and 'spirit,' and notions like 'beyond space and time.' In fact, even if anti-matter and matter prove to be completely unrelated, notions of the 'anti-physical' would be illegitimate. It's for this reason that double standards about physical theories are quite puzzling, as it 'is often said that quantum theory is deeply counterintuitive – e.g., in its description of the wave-like and particle-like behavior of fundamental particles, but no one seems to find it puzzling to suppose that it deals wholly with physical phenomena.'[203] If he were alive, Wittgenstein might ask 'What would it have looked like if it had looked as though quantum physics were true?'

Physics could, on the other hand, be 'finished,' in the sense that microphysics 'is already empirically adequate' and its 'ontology of substances' understood, as William G. Lycan has written. It could be that the only remaining ventures are 'positing superstrings, conducting a unified field theory and the like', which are 'only matters of interpreting and mathematizing the physical ontology.'[204] For Russell, 'Physics is mathematical, not because we know so much about the physical world, but because we know so little: it is only its mathematical properties that we can discover. For the rest, our knowledge is negative.'[205] Bringing our intuitions up to date with naturalistic inquiry, Giovanni Vignale writes that theoretical physics is

> the best compromise between escaping reality and diving into it. In mathematics you know exactly what you are talking about, because you have created the objects of your study. The non-existence of the perfect platonic circle is the best guarantee of its existence as an abstract object of study. In theoretical physics, however, you never know exactly what you are talking about: you are creating a fictional world to model a real world that you don't understand.[206]

Gödel's incompleteness theorems demonstrated too that certain mathematical systems cannot prove themselves to be true, forcing mathematics back into its 'mystic obscurity' with its users relying on inherently limited and unprovable axioms. In his *Mysticism and Logic* Russell suggested that 'Mathematics may be defined as the subject in which we never know what we are talking about, nor whether what we are saying is true.'[207]

And so there is, as the theoretical biologists stressed earlier, no alternative to what Husserl called the 'Galilean style' of physics; that is, constructing 'abstract mathematical models of the universe to which at least the physicists give a higher degree of reality than they accord the ordinary world of sensation.'[208] Pierre Jacob believes that 'linguistics and cognitive science should aim, in the Galilean style, for the same explanatory depth as theoretical physics: the goal of linguistics and cognitive science should be to make the appropriate idealizations that will lead to the discovery of unexpected principles that will be as removed from the empirical evidence, as are physical laws.'[209] Today, the study of the linguistic and mental aspects of the world are slowly leaving philosophy and integrating themselves into cognitive science, similar to how the philosophy of celestial motion was integrated into physics and astronomy. Consequently, the words of Stephen Weinberg are of even greater importance today than they were in 1976:

> The universe does not seem to have been prepared with human beings in mind, and the idea that humans can build mathematical models of the universe and find that they work is remarkable. Of course, one may doubt that the Galilean style will continue to be successful; one may question whether the mathematical ability of human beings can penetrate to the level of the laws of nature. To recall a statement by J. B. S. Haldane in which he said, in effect, that the universe is not only a good deal queerer than we know; it is a good deal queerer than we can know. Sometimes I believe that this is true. But suppose it is not. Suppose that by pursuing physics in the Galilean style we ultimately come to an understanding of the laws of nature, of the roots of the chains of explanation of the natural world. That would be the queerest thing of all.[210]

Russell positions himself close to Weinberg by claiming in 1912 that 'the truth about physical objects *must* be strange.'[211] So it is not at all controversial, as Chomsky notes, that 'science does not try to capture the content of ordinary discourse, let alone more creative acts of imagination. Paraphrasing Nagel, we cannot "find a place in the world" of physics for physical phenomena, as we describe them in physicalistic talk.'[212] It is of no surprise that the same is true of mental phenomena. For Hume, Newton's most important achievement was that while he 'seemed to draw the veil from some of the mysteries of nature, he shewed at the same time the imperfections of the mechanical philosophy; and thereby restored [Nature's] ultimate secrets to that obscurity, in which they ever did and ever will remain.'[213] Newton also agreed with many of his contemporaries who thought the notion of action at a distance was ridiculous and 'inconceivable,' since it is 'so great an Absurdity, that I believe no man who has in philosophical matters a competent Faculty of thinking, can ever fall into it.'[214] Scientists have, since Newton, abandoned the hope of trying to explain phenomena (observable occurrences or 'events'). Locke was also perceptive when, echoing Newton's 'spooky action' of gravity, he argued (countering the popular physicalist trend of denying the existence of free will) that motion has effects 'which we can in no way conceive motion able to produce':

> I deny not but a man, accustomed to rational and regular experiments, shall be able to see further into the nature of bodies and guess righter at their yet unknown properties than one that is

> a stranger to them; but yet, as I have, this is but judgment and opinion, not knowledge and certainty. This way of getting and improving our knowledge in substances only by experience and history, which is all that the weakness of our faculties in this state of mediocrity which we are in in this world can attain to, makes me suspect that natural philosophy is not capable of being made a science. We are able, I imagine, to reach very little general knowledge concerning the species of bodies and their several properties.[215]
>
> The idea of the beginning of motion we have only from reflection on what passes in ourselves, where we find by experience, that barely by willing it, barely by a thought of the mind, we can move the parts of our bodies, which were before at rest.[216]

This is something Rorty was aware of: 'Galileo and his followers discovered, and subsequent centuries have amply confirmed, that you get a much better prediction by thinking of things as masses of particles blindly bumping into each other than by thinking of them as Aristotle thought of them – animistically, teleologically, and anthropomorphically.'[217] But even this view presupposes the existence of only random or determinant entities, with nothing else (like the creative aspect of language or Locke's 'motion') lying beyond it. Writing on Democritus, Gerald Feinberg's words depict the approach of the pre-Socratic atomists as an anticipation of Newton's methods:

> What is remarkable is that he was willing to make the intellectual leaps of assuming the existence of unobserved objects quite different from those found in ordinary matter, and to account for everyday objects in terms of them. It is in this sense that Democritus is a forerunner of modern physics, in which the properties of bulk matter are accounted for in terms of atoms and their component particles, which in themselves behave very differently from the way bulk matter does.[218]

But it was Newton, not Democritus, who made the prominent excuses for dualism redundant, as Friedrich Lange explained:

> We have in our own days so accustomed ourselves to the abstract notion of forces, or rather to a notion hovering in a mystic obscurity between abstraction and concrete comprehension, that we no longer find any difficulty in making one particle of matter act upon another without immediate contact. We may, indeed, imagine that in the proposition, "No force without matter," we have uttered something very

Materialistic, while all the time we calmly allow particles of matter to act upon each other through void space without any material link.[219]

Aldous Huxley, promoting Galilean inquiry, also reviewed how 'The physical sciences started to make progress when investigators shifted their attention from qualities to quantities, from the appearance of things perceived as wholes to their fine structures; from the phenomena presented to consciousness by the senses to their invisible and intangible components, whose existence could only be informed by analytical reason.'[220] It's for these reasons that the phrase 'Theory of Everything' in physics is quite misleading, falling necessarily short of its grand title. Long before Newton, Socrates had 'expressed surprise that it was not obvious to them that human minds cannot discover these secrets, inasmuch as those who claim most confidently to pronounce upon them do not hold the same theories, but disagree with one another like lunatics.'[221] Socrates' distrust of the physical sciences should not be something to hold on to, but his pronouncement of the ultimately 'secret' nature of the universe cannot be easily dismissed. Wittgenstein, like Socrates, in his *Philosophical Investigations* thought that 'The existence of the experimental method makes us think we have the means of solving the problems which trouble us; though problem and method pass one another by.'[222] According to Xenophon, Socrates 'did not discourse about the nature of the physical universe, as most other philosophers did, inquiring into the constitution of the cosmos (as the sages call it) and the causes the causes of the various celestial phenomena; on the contrary, he pointed out the foolishness of those who concerned themselves with such questions. In the first place, he inquired whether they proceeded to those studies only when they thought they had a sufficient knowledge of human problems, or whether they felt that they were right in disregarding human problems and inquiring into divine matters.'[223] One can see Russell's approval of this method of understanding reality in many of his writings, or when he tells us that 'A prudent man imbued with the scientific spirit will not claim that his present beliefs are wholly true, though he may console himself with the thought that his earlier beliefs were perhaps not wholly false.'[224]

Perhaps Peter Atkins' assumptions of scientific progression seem welcome amongst these views: 'When we have dealt with the values of the fundamental constants by seeing that they are unavoidably so, and have dismissed them as irrelevant, we shall have arrived at complete understanding. Fundamental science then can rest. We are almost there. Complete knowledge is just within our grasp. Comprehension is moving across the face of the Earth, like the sunrise.'[225] Though even this ode to Reason (later to be recapitulated in Atkins' most recent elegy *On Being*) falls prey to the dogma that the modern scientific revolutions have afforded humans and unlimited grasp on nature. Echoing Locke's description of billiard balls, Galen Strawson also explains:

> I may also feel I understand – see – why this billiard ball does *this* when struck in this way by that billiard ball. But in this case there is already a more accessible sense in which I don't really *understand* what is going on, and it is an old point that if I were to ask for and receive an explanation, in terms of impact and energy transfer, starting a series of questions and answers that would have to end with a reply that was not an explanation but rather had the form "Well, that's just the way things are."[226]

Likewise, Thomas Kuhn suggests that

> It does not, I think, misrepresent Newton's intentions as a scientist to maintain that he wished to write a *Principles of Philosophy* like Descartes but that his inability to explain gravity forced him to restrict his subject to the *Mathematical Principles of Natural Philosophy*. Both the similarity and the difference of titles are significant. Newton seems to have considered his magnum opus, the *Principia*, incomplete. It contained only a mathematical description of gravity. Unlike Descartes' *Principles* it did not even pretend to explain why the universe runs as it does.[227]

Another possibility to take into account is that the brain's architecture might, very plausibly, have evolved to its peak structure in terms of processing power. Prominent neuroscientists have drawn the conclusion that if the brain were to develop to a larger size, the channelling or 'firing' of information and neurons would decrease in rate. Making the neurons and axons any thinner than their current size, as a result of a larger brain or in order to increase the range of connections, would stretch them to breaking point.[228] Perhaps (or rather, undoubtedly), 'like all the others, we can imagine only so far as our own nervous system allows us?'[229] One way to speed up our brain, adds David Robson, 'would be to evolve neurons that can fire more times per second. But to support a 10-fold increase in the 'clock speed' of our neurons, our brain would need to burn energy at the same rate as Usain Bolt's legs during a 100-metre sprint.'[230] John Duncan, in his book *How Intelligent Happens*, takes an entire three pages to describe the active neural network in the brain of a toad when it sees, and reacts to, a worm. When it comes to humans, however, the desire to map all cognitive capacities to accurate neural activity is a goal which current understanding and technology prohibits (as the nematode case also neatly demonstrates). Dennett adds, complimenting Duncan:

> Nervous systems that are hard-wired are lightweight, energy-efficient, and fine for organisms that cope with stereotypic environments on a limited budget. Fancier brains, thanks to their plasticity, are capable not just of stereotypic anticipation, but also of adjusting to trends. Even the lowly toad has some small degree of freedom in how it responds to novelty, slowly altering its patterns of activity to track – with considerable time lag – these changes in features of its environment that matter most to its wellbeing.[231]

Freeman Dyson's *Disturbing the Universe* gives a corresponding interpretation of modern physics:

> This experience brought home to me as nothing else could the truth of Einstein's remark, "One may say the eternal mystery of the world is its comprehensibility." Here was I, sitting at my desk for weeks on end, doing the most elaborate and sophisticated calculation to figure out how an electron should behave. And here was the electron on my little oil drop, knowing quite well how to behave without waiting for the result of my calculation. How could one seriously believe that the electron really cared about my calculation, one way or the other? And yet the experiments at Colombia showed that it did care. Somehow or other, all this complicated mathematics that I was scribbling established rules that the electron on the oil drop was bound to follow. We know that this is so. Why it is so, why the electron pays attention to our mathematics, is a mystery that even Einstein could not fathom.[232]

Since Einstein's time, physics has revealed that the universe is fundamentally a positively charged quantum void, and it was the difference between the amounts of matter and anti-matter which allowed it to come into existence to begin with – why there is something rather than nothing. Contemporary physics holds that entropy, which measures the number of ways a system can be rearranged, always tends to increase, in accordance with the second laws of thermodynamics (those laws which Lisa Simpson broke in creating her own perpetual motion machine, which infuriated her father: 'In this house we obey the laws of thermodynamics!'). The molecules of a hot gas can easily be rearranged to leave the overall temperature intact, whereas the molecules in a complex life form don't have as much freedom, since mass rearrangement would bring the organism to an early demise. 'By the same logic,' writes Amanda Gefter, 'nothingness is the highest entropy state around – you can shuffle it around all you want and it still looks like nothing.'[233] Because of this, it's hard for the physicist to explain how something could arise from nothing (or even if this proposition has any meaning).

It's here that we return to symmetry, since for the physicist nothingness is entirely symmetrical. 'There's no telling one part from another, so it has total symmetry,' according to Frank Wilczek, a physicists working in quantum chromodynamics, the field which describes the activity of quarks within atomic nuclei. Gefter believes the theory tells us that 'nothingness is a precarious state of affairs' since, as Wilczek adds, 'You can form a state that has no quarks and antiquarks in it, and it's totally unstable. It spontaneously starts producing quark-antiquark pairs' – breaking the symmetry of nothingness.[234]

This has lead Victor Stenger to conclude that, in spite of entropy, 'something is the more natural state than nothing.' Gefter goes on to report that 'Emptiness would have precisely zero energy, far too exacting a requirement for the uncertain quantum world. Instead, a vacuum is actually filled with a roiling broth of particles that pop in and out of existence.'[235] This could be an explanation behind the origins of the universe: 'There is no barrier between nothing and a rich universe full of matter,' thinks Wilczek, to which Gefter adds: 'Perhaps the big bang was just nothingness doing what comes naturally.' But as Stephen Hawking has warned, asking what came before the Big Bang is like asking what's north of the North Pole. But 'there is,' adds Gefter, 'an even more mind-blowing consequence of the idea that something can come from nothing: perhaps nothingness itself cannot exist.' A more precise reason why 'nothingness' doesn't exist is that

> Quantum uncertainty allows a trade-off between time and energy, so something that lasts a long time must have little energy. ... That fits with the generally accepted view of the universe's early moments, which sees space-time undergoing a brief burst of expansion immediately after the big bang. This heady period, known as inflation, flooded the universe with energy. But according to Einstein's general theory of relativity, more space-time also means more gravity. Gravity's attractive pull represents negative energy that can cancel out inflation's positive energy – essentially constructing a cosmos for nothing. ... Physicists used to worry that creating something from nothing would violate all sorts of physical laws such as the conservation of energy. But if there is zero overall energy to conserve, the problem evaporates – and a universe that simply popped out of nothing becomes not just plausible, but probable. None of this

really gets us off the hook, however. Our understanding of creation relies on the validity of the laws of physics, particularly quantum uncertainty. But that implies that the laws of physics were somehow encoded into the fabric of our universe before it existed. How can physical laws exist outside of space and time and without a cause of their own? Or, to put it another way, why is there something rather than nothing?[236]

It wouldn't be a submission of failure to do as Christiaan Huygens did in 1698, and admit 'I shall be very well contented, and shall count I have done a great matter, if I can but come to any knowlege of the nature of things, as they now are, never troubling my head about their beginning, or how they were made, knowing that to be out of the reach of human Knowlege, or even Conjecture.'[237] Or as Michio Kaku put it: 'our bodies are essentially symphonies. They are made out of vibrating strings. The universe obeys the laws of physics, and the laws of physics are nothing but the laws of harmony.'[238] This sounds reminiscent of Kepler's astronomical theory in 1619 work *The Harmonies of the World*, which 'affirmed that the planet's elliptical orbits caused each to produce a series of rising and falling notes, radically unlike the supposed monotonous droning of the Ptolomaic spheres; together, the planets sang in a polyphony that could be heard only by the Composer.'[239] This added to the beliefs of other pre-Newtonian astronomers, who thought the planets of the solar system had senses. This is similar to Schopenhauer's view of astronomy, 'where heavenly bodies sport with each other, betray inclination, and as it were exchange amorous glances, though never driving matters so far as coarse contact, but, keeping due distance, decorously dance their minuet to the music of the spheres.'[240]

With implications for the search in biolinguistics for 'third factor' effects on language, Pierre Jacob points out that one of the 'internalist principles' Chomsky appeals to

> is a principle of symmetry (or parallelism), which (in accordance with his own minimalist program) governs the computational architecture of the grammars of natural languages. According to this principle of symmetry, the syntax of an I-language generates mental representations on which the rules of phonological and semantic interpretation operate in parallel. The mental representations generated by syntax thus constitute a twofold set of instructions for both the human sensorimotor system (which controls the articulation and perception of the sounds of language) and the human conceptual system (which controls interferences).[241]

Humanity, for V. S. Ramachandran, 'transcends apehood to the same degree by which life transcends mundane chemistry and physics.'[242] An astonishing conclusion of the bourgeoning discipline of neuroscience is that there are more neurological connections in the human brain than there are particles in the known universe. As Douglas Fox understands, 'With 100 billion neurons, each with up to 10,000 connections, or synapses, the human brain is the most complex object in the known universe.'[243] Ramachandran continues by proposing that 'The question of how neurons encode meaning and evoke all the semantic associations of an object is the holy grail of neuroscience, whether you are studying memory, perception, art, or consciousness.'[244] The 'Gatsby Unit' of Computational and Theoretical Neuroscience and Machine Learning at UCL considers the brain to be 'perhaps the most complex subject of empirical investigation in scientific history.'[245] It's of no surprise that for Gerald M. Edelman, winner of the 1972 Nobel Prize for Physiology, 'the methods of science that are used to study inanimate objects are not up to the task of doing science on animals that have brains and possess intentionality.'[246]

Putting aside the concerns of quantum theorists, neuroscientist John Duncan suggests that 'the full account of how human thought emerges from a biological brain, a network of billions of neurons communicating via tiny electrical impulses, still ranks among the great scientific mysteries.'[247] Like Ramachandran, Jaegwon Kim writes in his introduction to the philosophy of mind: 'How meaning and understanding could arise out of molecules and cells is as much a mystery as how they could arise out of strings of 0s and 1s.'[248] How neurons encode information is at the foundation of these basic operations, and presently eludes explanation. The current era of neuroscience is yet to correct C. R. Gallistel, who wrote in 1997 that 'we clearly do not understand how the nervous system computes' even for 'the small scale set of arithmetic and logical operations that are fundamental to any computation.'[249] Gallistel was reviewing the computations of ant colonies, and extending this watermark to the human brain reveals the striking gulf between neurology and psychology. Along with this comes the conclusion that 'the neurophysiology of language,' one of the most complex faculties of the mind, 'remains almost a total mystery.' We lack an understanding of how to relate the cognitive theories of the rules and representations of language to the atomic, cellular, and neurological properties of the brain.[250]

It might also be discovered, not unreasonably, that 'no explanation solely in terms of brain processes will be such that we can deduce the existence of consciousness from it,' meaning that 'someone could know all the physical facts about the world and still not know about consciousness.'[251] Hauser qualifies that we also currently lack an

> account of how the brain of a bee creates the representations of its own language, what von Frisch described in the 1960s as the bee's dance language. We certainly know that the dance is, in some respects, symbolic, in that it stands for or provides information about the location and quality of food. But we don't know how electrical activity creates this information in a format that can be read out and followed. And I don't think we are even close to understanding this problem. Scale it up, and ask how the brain creates the representations that enable us to appreciate the grammaticality of colorless green ideas sleep furiously, while also appreciating the lack of intelligible meaning, and we come up remarkably short.[252]

Timothy Wilson and Elizabeth Dunn have also concluded that introspection 'does not provide a direct pipeline to nonconscious mental processes. Instead, it is best thought of as a process whereby people use the contents of consciousness to construct a personal narrative that may or may not correspond to their nonconscious states.'[253] Introspection allows certain access to conscious feelings (content) but mental processes remain hidden – contrary to Leibniz's conviction that they 'can be discovered in us by dint of attention.'

It's important to note that the 'starting line,' in Strawson's words, has not even been reached if one does not appreciate the thought that consciousness is a purely physical phenomenon – if this idea is rejected, one is not a materialist in any meaningful sense of the term. It's perhaps unfortunate that our language attaches metaphysical connotations to the word 'property.' Thinking that 'consciousness is a property of the brain' makes it seem as if consciousness is something 'beyond' the physical brain. A 'belief' in 'real' 'materialism' may therefore be easier to grasp in Eastern schools of meditation:

> It can help to perform special acts of concentration – focusing one's thought on one's brain and trying to hold fully in mind the idea that one's experience as one does so is part of the physical being of the brain (part of the physical being of the brain that one may be said to be acquainted with as it is in itself, at least in part, because its being as it is for one as one has it just is what it is in itself, at least in part). It is worth trying to sustain this – it is part of doing philosophy – forcing one's thought back to the confrontation when it slips. At first one may simply encounter the curious phenomenological character of the act of concentration, but it is useful to go on – to engage, for example, in silent, understanding – engaging subvocalizations of such

thoughts as "I am now thinking about my brain, and am thinking that this experience I am now having of this very thinking – and this subvocalization – is part of the physical activity and being of my brain." It is also useful to look at others, including young children, as they experience the world, and to think of the common-or-garden matter that is in their heads (hydrogen, oxygen, carbon, iron, potassium, sodium, and so on). It is useful to listen to music, and focus on the thought that one's auditory experience is a form of matter.[254]

Kant thought along similar lines: 'we intuit ourselves only as we are inwardly affected *by ourselves*; and this would seem to be contradictory, since we should then have to be in a passive relation [of active affection] to ourselves. It is to avoid this contradiction that in systems of psychology *inner sense*, which we have carefully distinguished from the faculty of *apperception*, is commonly regarded as being identical with it.'[255] Here is Heisenberg's interpretation of the physics of his day and the implications for the interpretation of the 'natural' world, along with Russell's view of the interaction between the emerging physics of the day and our 'folk science':

> Modern science shows us that we can no longer regard the building blocks of matter, which were considered originally to be the ultimate objective reality, as being things 'in themselves' ... Knowledge of atoms and their movements "in themselves", that is to say independent of our observation, is no longer the aim to research; rather we now find ourselves from the very start in the midst of a dialogue between nature and man, a dialogue of which science is only one part, so much so that the conventional division of the world into subject and object, into inner world and outer world, into body and soul, is no longer applicable and raises difficulties. For the science of nature, the subject matter of research is no longer nature in itself, but nature subjected to human questioning, and to this extent man, once again, meets only with himself.[256]

> We all start from "naive realism," i.e., the doctrine that things are what they seem. We think that grass is green, that stones are hard, and that snow is cold. But physics assures us that the greenness of grass, the hardness of stones, and the coldness of snow are not the greenness, hardness, and coldness that we know in our experience, but something very different. The observer, when he seems to himself to be observing a stone, is really, if physics is to be believed, observing the effects of the stone upon himself. Thus science seems to be at war with itself: when it most means to be objective, it finds itself plunged into subjectivity against its will. Naive realism leads to physics, and

physics, if true, shows that naive realism is false. Therefore naive realism, if true, is false; therefore it is false.[257]

Russell's observations of perception are clarified by Rodolfo Llinás, for whom 'seeing is reconstructing the external world, based not on the reflecting properties of light on external objects but, rather, on the transformation of such visual sensory input (a vector) into perception vectors in *other sets of coordinate systems*.'[258] In his 1963 essay 'Observation and the Will,' the Australian philosopher Brian O'Shaughnessy explained (echoing William James) that 'The astonishing thing about action is that it is possible at all. For, if a man is making a chair, you will find a physical causal explanation of the movement of each piece of wood from its initial to its final setting; everything that happens is in accordance with law; but you will look throughout this world or universe forever in vain for an analogous physical explanation of their coming together in the form they did, a form that mirrors human need and the human body itself. (Try it.)'[259] Samuel Guttenplan is of the opinion that this shows how 'you couldn't imagine the pieces coming together unless there was some mind *orchestrating* the movement. But the action itself – the physical movements of the pieces – is not mental. What happens is that you see these movements – the actions – and then *infer* that there are mental states directing them. In seeing the actions, you don't literally *see* the mind.'[260]

Chomsky once commented, in an interview with Bryan Magee, that 'As soon as questions of will or decision or reason or choice of action arise, human science is at a loss.'[261] It's for this reason that ethnoscience (to choose a new definition from Milena Nuti's doctrinal thesis: the exploration of the cognitive basis for the intuitive and 'untutored ideas people have about the world and how it works') and the humanities will continue to be a necessary path to any decent level of human fulfilment, perhaps in turn being able to avoid the fate of Darwin who, after re-reading Shakespeare in his later years, was overcome with a boredom so intense he felt physically sick.[262]

The task remains a difficult one, ever since the seventeenth-century 'natural philosophers' decided to abandon a mechanistic interpretation of the world – leading Newton to conclude that when an object falls it does so not because, as the theologians and Aristotelians believed, it was falling to its natural place, but because it was obeying the 'absurdity' of action at a distance. 'There has been very valuable work,' adds Chomsky, 'about how an organism executes a plan for integrated motor action – say, how a person reaches for a cup on the table. But no one even raises the question of why this plan is executed rather than some other one, apart from the very simplest organisms and special circumstances of motivation.'[263] Thomas Huxley, in 1854, spoke differently to Chomsky, but a similar point was still made:

> When Newton saw the apple fall, he concluded at once that the act of falling was not the result of any power inherent in the apple, but that it was the result of the actions of something else on the apple. In a similar manner, all physical force is regarded as the disturbance of an equilibrium to which things tended before its exertion, – to which they will tend again after its cessation. ... But to the student of Life the aspect of Nature is reversed. Here, incessant, and, so far as we know, spontaneous change is the rule, rest the exception – the anomaly to be accounted for. Living things have no inertia, and tend to no equilibrium.[264]

In his 'Poem on the Lisbon Disaster,' based on the 1755 Lisbon Earthquake, Voltaire posed questions in a footnote which still, to this day, we cannot answer with complete confidence: 'How is our brain capable of ideas and memory? In what manner do our limbs obey every motion of the will? Of all this we are entirely ignorant.'[265] In his chapter on 'Physiology and Pathology' in *The Will in Nature*, Schopenhauer concurs that 'perceivance and thinking will be ever more and more explained from the organism, but willing never.'[266] Amongst others, Rousseau believed that the power of choice of action and its 'sentiments' were beyond physical explanation:

> Nature commands every animal, and the beast obeys. Man feels the same impetus, but he realizes that he is free to acquiesce or resist; and it is above all in the consciousness of this freedom that the spirituality of his soul is shown. For physics explains in some way the mechanism of the senses and the formation of ideas; but in the power of willing, or rather of choosing, and in the sentiment of this power are found only purely spiritual acts about which the laws of mechanics explain nothing.[267]

Georges Rey writes that intentionality is 'peculiarly resistant to incorporation in the rest of science.'[268] Many philosophers and biologists believe the individual's *umwelt* (self-centred world; worldview; any meaningful aspects of the world for a particular organism, such as concern for food, shelter, potential threats, mating partners, points of reference or navigation, and so on) cannot be 'pinned down' by science: 'We understand others, but we cannot reduce this understanding to a branch of the natural sciences.'[269] Part of our *umwelt* is based on a theory that other minds exist, but the problem of how this particular knowledge is represented in the brain in terms of neurophysiology 'has not been solved.' Instead, 'the subject has undergone a sort of naturalization.'[270]

Similar to intentionality, William James focused the study of mind on conscious attention: 'Everyone knows what attention is. It is the taking possession by the mind, in clear and vivid form, of one out of what seem several simultaneously possible objects or trains of thought.'[271] The primary sensory tool of attention is vision, something which Da Vinci was curious of: 'The eye, which is called the window of the soul, is the chief means whereby the understanding may most fully and abundantly appreciate the infinite works of nature.'[272] The extreme importance we place on the visual system is without question, but many of its dynamic functions still go without explanation – let alone the questions of will and choice which underlie its operation.

The German physicist Hermann von Helmholtz noted how we can shift the focus of our attention between multiple objects whilst keeping our eyes at a fixed position within the same visual field. Since Helmholtz's time in the nineteenth century, there have been various developments into the nature of this phenomenon, but no progress into the will or choice behind it.[273] We are reminded by a prominent neuroscientist that 'how the brain combines the responses of these cells to indicate a continuous vertical line is a mystery that neurology has not yet solved.'[274] In addition, it's still unknown how one continuous line is distinguished from others in the same visual field, lending some satisfaction to Delacroix's appeal that 'it would be worthy to investigate whether straight lines exist only in our brains.'[275]

During a discussion on Wittgenstein's duck-rabbit picture (a drawing which looks both like a rabbit and a duck, depending on the individual's 'take'), Jessica Griggs noted that the 'precise neural mechanism that provokes the brain to switch its view of a scene is unknown, but it is thought to play a major role in perception by acting as a sort of reality check.'[276] Rudolf Arnheim's words in 1969 are still as important today: 'What we need to acknowledge is that perceptual and pictorial shapes are not only translations of thought products but the very flesh and blood of thinking itself.'[277] The areas of the brain which have a major influence on this intriguing phenomenon are the superior parietal lobes. 'Evolution never stops,' adds David Robson, who cites a recent study which found that 'the visual cortex has grown larger in people who migrated from Africa to northern latitudes, perhaps to help make up for the dimmer light up there.'[278]

Shimon Ullman's rigidity principle of vision stipulates that when 'Presented with the task of inferring structure from motion, our visual system works on the assumption that what is seen is a rigid body in movement rather than a static entity that is changing shape.'[279] As

Ullman himself put it, 'Computational studies have shown that the 3-D structure of rigid and quasirigid objects can be recovered by looking for the most rigid interpretation possible of the changing image,' structuring and limiting objecthood.[280] These findings help ease both of Locke's primary and secondary qualities into the category of mental constructions, contrary to his belief that the primary qualities of shape, mass and motion are mind-independent. Ullman's rigidity principle, as outlined by Donald Hoffman, stipulates that 'if possible, and other rules permit, interpret image motions as projections of rigid motions in three dimensions.'[281] The twentieth century's cognitive revolution was the revival of the examination of the nature and character of similar unconscious mental principles in the seventeenth and eighteenth centuries, going far beyond the likes of Freud. The unconscious syntactical processes of language, for instance, were merely examined at a descriptive level by the continental philosophers like Merleau-Ponty, whose work bears as much scientific legitimacy to linguistics as a short story about flying cars has to physics: 'The wonderful thing about language is that is promotes its own oblivion: my eyes follow the lines on the paper, and from the moment I am caught up in their meaning, I lose sight of them. The paper, the letters on it, my eyes and body are there only as the minimum setting of some invisible operation.'[282]

It's not too hard to swallow, then, the claim that the entirety of sensory perception 'is a figment of your imagination.'[283] In the famous Ponzo illusion, for instance, in which horizontal parallel lines overlap with vertical intersecting diagonal lines (forming what appears to be something like a receding train track), the perceived size disparity 'arises from the attempt on the part of the perceptual system to maintain size constancy across the entire visual array.'[284] Ivo Kohler and Warren Wittreich comment on similar issues:

> [T]he [visual] image is 'better' than it should be, considering the known defects in the visual system. For example, the lens of the eye is not corrected for spherical aberration; hence straight lines should look slightly curved. By the same token, lines of a certain curvature should appear straight. It is well known that the eye is not corrected for colour; as a result, different wavelengths of light – originating at a common point – do not come to a focus on the retina. One would expect this defect, called chromatic aberration, to have a noticeable effect on vision, but it does not, except under special conditions.[285]
>
> When we watch a person walk away from us, his image shrinks in size. But since we know for a fact that he is not shrinking, we make an unconscious correction, and 'see' him as retaining his

full stature. Past experience tells us what his true stature is with respect to our own. Any sane and dependable expectation of the future requires that he have the same true stature when we next encounter him. Our perception is thus a prediction; it embraces the past and the future as well as the present.[286]

Our sense of vision, often thought of as a passive phenomenon, has been proven to be a highly reflexive relationship between the eyes and the external world – we bring with us as much to the scene than what we take away. 'It is widely held,' says Dennett, 'that human vision, for instance, cannot be explained as an entirely 'data-driven' or 'bottom-up' process, but needs, at the highest levels, to be supplemented by a few 'explanation-driven' rounds of hypothesis testing (or something analogous to hypothesis testing).'[287] Physicists David Bohm and David Peat write how 'scientists have shown that seeing requires the active movement of both the body and the mind. Visual perception is therefore an intentional and not a passive act.'[288] Semir Zeki writes in classic work that the brain 'extracts from the continually changing information reaching it only that which is necessary for it to identify the characteristic properties of what it views; it has to extract constant features in order to be able to obtain knowledge about them and to categorise them.' Vision depends 'as much upon the operations of the brain as upon the external, physical, environment; the brain must discount much of the information reaching it, select from that information only that which is necessary for it to be able to obtain knowledge about the visual world and compare the selected information with its stored record of all that is seen.' Mounting evidence suggests, wrote Zeki, that the areas of the brain responsible for visual processing

> are also perceptual systems in that activity in each can result in a percept without reference to the other systems; each processing-perceptual system terminates its perceptual task and reaches its perceptual end-point at a slightly different time from others, thus leading to a perceptual asynchrony in vision – colour is seen before form which is seen before motion, the advantage of colour over motion being of the order of 60-100ms. Thus visual perception is also modular. In summary, the visual brain is characterised by a set of parallel processing-perceptual systems and a temporal hierarchy in visual perception.[289]

Eric Kandel, whose thinking is probably closest to Zeki's, sees the brain sciences as potentially being able to build important bridges between the sciences and the humanities, with the collaboration between visual scientists and art historians becoming a real possibility later in the century. James McGilvray also points out that vision 'seems to be far less in our control than language. As some 17th century philosophers put it, when we produce a colored visual scene in imagination, is it "less vivid" than those produced by sensory system promptings.'[290] We conventionally view our eyes, writes McGilchrist,

> like the lens of a camera of a moving swivel, perhaps a bit like a film-maker's camera ... The image suggests that we choose where we point our attention; in that respect we see ourselves as supremely active, and self-determining. As to the "impressions" we receive, we are like a photographic plate, taking a faithful record of the world "out there"; and in that we pride ourselves on objectivity, being supremely passive ... [In reality] [w]e know that we are neither as active in choosing where we direct our attention, nor as passive in the process of seeing, as this account suggests. There is another story to be told about seeing, and it is one that is better supported by neuroscience. It is also more in keeping with the right hemisphere's view of the world. According to this view, we are already in a relationship with the world, which helps to direct our attention; and which also means that we bring something of ourselves to the process of creating a "vision" of the world.[291]

In 1996, V. L. Deglin and Marcel Kinsbourne used electroconvulsive therapy to disable one of each the brain's hemispheres in a series of subjects to test their reasoning skills. All subjects had one hemisphere disabled and were asked to respond to a syllogism with a false premise, before repeating the process with the other hemisphere. The major premise in the syllogism was 'All monkeys climb trees.' The minor premise was 'The porcupine is a monkey.' The implied conclusion was 'The porcupine climbs trees.' But each hemisphere approached the truth of the deduction in different ways. 'At the outset of their experiment, when the intact individual is asked 'Does the porcupine climb trees?', she replies (using, of course, both hemispheres): 'It does not climb, the porcupine runs on the ground; it's prickly, it's not a monkey.' ... During experimental temporary hemisphere inactivations, the left hemisphere *of the very same individual* (with the right hemisphere inactivated) replies that the conclusion is true: 'the porcupine climbs trees since it is a monkey.' When the experimenter asks, 'But is the porcupine a monkey?', she replies that she knows it is not. When the syllogism is presented again, however, she is little nonplussed, but replies in the affirmative, since 'That's what is written on the card.' When the right hemisphere *of the same individual* (with the left hemisphere inactivated) is asked if the syllogism is true, she replies: 'How can it climb trees – it's not a monkey, it's wrong here!' If the experimenter points out that the conclusion must follow from the premises stated, she replies indignantly: 'But the porcupine is not a monkey!'"[292]

With various other false syllogisms, the same pattern between the approaches of the hemispheres has emerged. Truth for the right hemisphere is not to be found within systems signs, but in the relationship between things in the world, which it recognises do not exist independently, within a system of logic, in a vacuum, but in a world of constant flux and changing circumstance. The 'neglect of context,' thought Dewey, 'is the greatest single disaster which philosophic thinking can incur.'[293] McGilchrist concludes that 'All rationality can do it provide internal consistency once the system is up and running.'[294] McGilchrist concludes that 'objectivity *requires* interpretation of what one finds, depends on imagination for achievement. Detachment has a deeply ambiguous nature.'[295] Likewise, Gerald Feinberg explained in the 1960s how 'The proper understanding of matter requires the imagination to invent entities not apparent in everyday phenomena. It is the enduring miracle of creative thought that the mind is equal to this task.'[296] Russell appears to have interpreted 'objects' in a similar manner to Descartes' famous description of candle wax:

> In the presence of my table I am acquainted with the sense-data that make up the appearance of my table – its colour, shape, hardness, smoothness, etc. ... The particular shade of colour that I am seeing may have many things said about it – I may say that it is brown, that it is brown, that it is rather dark, and so on. But such statements, though they make me know truths *about* the colour, do not make me know the colour itself any better than I did before: so far as concerns knowledge of the colour itself, as opposed to knowledge of truths about it, I know the colour perfectly and completely when I see it, and no further knowledge of it itself is even theoretically possible. Thus the sense-data which make up the appearance o my table are things with which I have acquaintance, things immediately known to be just as they are.[297]

In his 1914 Lowell Lectures, Russell expanded his ideas on how things appear to us 'just as they are': 'It is a mistake to speak as if acquaintance had degrees: there is merely acquaintance and non-acquaintance. When we speak of becoming "better acquainted," as for instance with a person, what we must mean is, becoming acquainted with more parts of a certain whole; but the acquaintance with each part if either complete or non-existent.'[298] He also strongly believed that if we treat ordinary objects in our world as logical fictions, 'they are extruded from the world of what there is, and in their place as what there is you find a number of passing particulars of the kind that one is immediately conscious of in sense. I want to make clear that I am not *denying* the existence of anything; I am only refusing to affirm it.'[299]

For Descartes, the infinite variety and creativity of language use was proof of the existence of other minds. This was most likely because a significant part of conscious thinking is language-based and separates us from all other species – it is 'the mark of mind.' More recently, Davidson has asked (in a manner of some relevance for the issue of artificial intelligence discussed above): 'How can we tell if a creature is thinking? I fear the only answer is too simple to seem philosophically interesting. If we can communicate with the creature on a range of topics concerning our shared environment, that creature is conscious and it is thinking. It is a little more interesting that there is no other way to tell.'[300] Though I wouldn't go as far to call it 'faith,' the core of G. K. Chesterton's following point seems correct and comparable to Davidson's: 'Reason is itself a matter of faith. It is an act of faith to assert that our thoughts have any relation to reality at all.'[301] Contra Dennett, 'One cannot even sensibly speak of an "illusion" in an unconscious sensory processing (error yes, but illusion no).'[302] Since the time of Newton (and more forcefully with the advent of quantum physics) science has been forced to fix its gaze on idealised aspects of the world (the chemical, optical, biological, and so on) and leave the interpretation of everyday phenomena to ethnoscience.

John Searle develops a different line of thought, whereby our inferences that others are conscious are not based solely on behaviour, but on the observation that they have the same 'causal structure' as us: Namely, 'they have eyes, nose, skin, mouth and all the rest, and the behaviour is relevant to the question of their conscious states only because we see it as situated in an overall causal order.'[303] It's also important to relate theories of creativity to the concepts of will, choice and intention, in the sense that it is not at all obvious 'that anyone will ever be able to explain the creative impulse, and it is unlikely that anyone would ever want to do so.'[304] Firm causal structure or not, the human body nevertheless decays and sees the vast majority (though not all, as is commonly claimed) of its cells replaced over long periods, and so the question of personal identity based on a consistent molecular basis becomes severely reduced in relevance. This is even illustrated in children's fairy tales, when the handsome knight is turned into a frog, kisses the princess, and reverts back (what's come to be known as psychic continuity). The child is aware throughout that the frog is the prince, despite bearing no resemblance to him. Such concepts have a form of topology, capable of being 'warped' but remain intact, as when

Looney Tunes characters are squashed flat before recovering. More complex examples can be drawn from science fiction or popular culture. In the video game *Crysis 2*, for instance, the character Prophet dies whilst wearing an advanced nanosuit, and a fellow soldier named Alcatraz finds his body, takes off the special suit and keep it for himself (having also been given the slight task of saving the world from an alien invasion). But the nanosuit has a special 'memory' system which (somehow) stores the mind of whoever wears it. Later on Alcatraz is killed, but the nanosuit uploads Prophet's mind into Alcatraz's body, bringing the first soldier, in a sense, back to life. Prophet is the same 'person,' and 'Alcatraz' no longer exists despite the fact that his 'body' is still alive. Switching abstract personal identities, as in the case of the prince and the frog, cannot be done with objects lacking in personal identity features: a radiator cannot trade identities with a bookmark. Limitations to psychic continuity are also imposed by noun types, since 'you can get Sylvester [the cartoon donkey] to turn into the wing but it's more difficult to get the wind to turn into Sylvester.'[305] 'To see the organism *in* nature,' wrote Dewey, 'the nervous system in the organism, the brain in the nervous system, the cortex in the brain is the answer to the problems which haunt philosophy. And when thus seen they will be seen to be *in*, not as marbles are in a box but as events are in history, in a moving, growing, never finished process.'[306]

The seventeenth century Dutch mathematician, Christiaan Huygens, wrote in his *Cosmotheoros* that 'if we do but consider some sorts of Beasts, as the Dog, the Ape, the Beaver, the Elephant, nay some Birds and Bees, what sense and Understanding they are masters of, we shall be forc'd to allow, that Man is not the only rational Animal. For we discover somewhat in them of Reason independent on, and prior to all teaching and practice.'[307] Various studies, incidentally, have also shown how certain animals have a 'theory of mind' and a sense of self (although frogs, alas, are not amongst their number).[308] Support for these theories include 'recent studies of chimpanzees [which] suggest that they recognize the perceptual act of seeing as a proxy for the mental state of knowing'[309] At the same time, 'other studies suggest that even chimpanzees lack a theory of mind, failing, for example, to differentiate between ignorant and knowledgeable individuals with respect to intentional communication.'[310] Barry C. Smith's entry in the *Oxford Handbook of Philosophy of Language* explains how there 'is evidence that many animals have numerosity and even some arithmetic but this is usually limited to between 4 and 6 items. Even monkeys taught to count up to 4 and do some subtraction and addition, do not naturally progress to 5 but have to be taught to deal with this new number all over again through repeated trials. They show no ability, of the sort the child has, to extend the series indefinitely; that is, they have no concept of successor,' lacking our recursive property of mind.[311]

They also appear to lack the concept of 'person,' an inherently abstract notion (as opposed to the concrete concept of a 'body'). To take a famous case of this principle in action, a man suffering from Capgras syndrome will think that his wife (or another significant person) has been replaced by an imposter who is physically identical to her; that is, he keeps the concrete identity of his wife, but loses the abstract one.[312] 'God' could also signify the abstract qualities of humanity (moral standards etc.) as opposed to the concrete ones – hence the confusion of most Catholic teenagers who are told that, through the doctrine of transubstantiation, Christ and the bread on the alter are the same 'thing.' McGilchrist claims that the

> right hemisphere, as one can tell from the fascinating changes that occur after unilateral brain damage, is responsible for our sense of the body as something we "live", something that is part of our identity, and which is, if I can put it that way, the phase of intersection between our selves and the world at large. For the left hemisphere, by contrast, the body is something from which we are relatively detached, a thing in the world, like other things (*en soi*, rather than *pour sui*, to use Sartre's terms), devitalised, a "corpse". As Gabriel Marcel puts it, it is sometimes as if I *am* my body, sometimes as if I *have* a body.[313]

This view has been shared by Michael Ayers, who writes that 'our experience of *ourselves* as being a material object among others essentially permeates our sensory experience of things in general.'[314] This sense of detachment leads John Muir to conclude: 'Most people are *on* the world, not in it – have no conscious sympathy or relationship to anything about them – undiffused, separate, and rigidly alone like marbles of polished stone, touching but separate.'[315] McGilchrist's final word on the matter leaves us questioning other deep-seated assumptions:

> Although it might seem that we overvalue the body and physical existence in general, that is not what I deduce from our preoccupation with exercise, health and diet, with "lifestyles," concerned though this is with the body and its needs and desires. Nor does it follow from the fact that the body was never so much on display, here or in cyberspace. The body has become a thing, a thing we possess, a mechanism, even if a mechanism for fun, a bit like a sport car with a smart sound system. That mechanistic view derives from the nineteenth-century scientific world picture, which has lingered with us longer in biology and

the life sciences that in physics. The body has become an object in the world like other objects, as Merleau-Ponty feared. The left hemisphere's world is ultimately narcissistic, in the sense that is sees the world "out there" as no more than a reflection of itself: the body becomes just the first thing we see out there, and we feel impelled to shape it to our sense of how it "should" be.[316]

These consumerist obsessions are spurred on by the public relations industry, which encourages people to 'perform leisure,' as political economist Thorstein Veblen put it.

This was an issue Locke spent a great deal of time on, concluding that 'personal identity' is directly connected to a stream of connected memories or consciousness. Locke's approach is in many ways like that of the modern cognitive scientist, taking personal identity as a 'forensic' matter, and 'who endeavours to prove empirically that there is no unique scene of the Self, just a pandemonium of competing forces.'[317] In an essay on the topic, Eric T. Olson comments:

> I believe that the Psychological Approach [based on Locke's writings] owes much of its popularity to the fact that philosophers typically begin their inquiries into personal identity by asking what it takes for us to persist through time. But an equally important question is what we are: whether we are animals, what we might be if we aren't animals, and how we relate to those animals that some call our bodies, for instance. This question is often ignored, or addressed only as an afterthought. That is why philosophers have failed to appreciate the problem of the thinking animal. Perhaps they ought instead to begin by asking what we are, and only then turn to our identity through time and other matters. Many would end up thinking differently.[318]

This argument echoes through to issues of 'minds' and 'bodies' and other pre-theoretical notions (and, in fact, our understanding of the world in general). It's with this in mind that we read Harold Noonan, who writes that 'personal identity is essentially no different from the identity of material objects in general,' though this ignores the particularly complex nature of a 'person' as opposed to a 'chair.'[319] Timothy D. Wilson outlines some of the 'key principles established by social psychological research':

> The self is an inherently social construct, shaped by one's culture, family, and peers.
> Introspection is severely limited. Much of the workings of the human mind, including attributes we generally think of as part of the 'self,' are unavailable to conscious scrutiny (note that this

contradicts the premise that the study of the self is the same as the study of consciousness).
The act of introspection can, depending on various well-known conditions, lead to long-term recovery from personal trauma or ill-advised choices that people later regret.
People observe their own behavior to deduce who they are.
The narratives people tell themselves are key determinants of their future behavior, regardless of whether these narratives are objectively true.[320]

Hume's conviction was that our sense of self is 'fictitious,' and simply a bundle of our experiences and cognitive capacities:

> For my part, when I enter most intimately into what I call *myself*, I always stumble on some particular perception or other, of heat or cold, light or shade, love or hatred, pain or pleasure. I never can catch *myself* at any time without a perception, and never can observe anything but the perception.[321]

We should also recall Hume's belief that "tis evident, the same method of reasoning must be continu'd which has so successfully explain'd the identity of plants, and animals, and ships, and houses, and of all the compounded and changeable productions either of art or nature. The identity, which we ascribe to the mind of man, is only a fictitious one, and of a like kind with that which we ascribe to vegetables and animal bodies. It cannot, therefore, have a different origin, but must proceed from a like operation of the imagination upon like objects.'[322] Contrary to the dominant view of him as the arch-empiricist, in the *Enquiry* Hume

> effectively suggests, strikingly enough, that externalist and empiricist doctrines, in the sense of both a belief in mind-independent objects of which we then form representations, and a faith in sense experience, is nothing but the expression of a "powerful instinct of nature", "infallible and irresistible", "a universal and primary opinion of all men". In other words, an empiricism and externalism of this sort is innate. Philosophy has to *correct* this basic instinct, which wrongly suggests to us the existence of external objects causing our representations of them.[323]

Prominent developments in neuroscience reveal that 'the frontal lobe of the brain may be where the self is; or an important aspect of it.'[324] Regardless of this, the taxonomical problems of the past prove that the 'development of the biological concept of the species is one of the earliest manifestations of the emancipation of biology from an inappropriate philosophy based on the phenomena of inanimate nature.'[325]

It's also been shown over the course of the last two decades that imagination and visual perception 'share important neural foundations.' It is at a clear neurological level that when we imagine doing something it is as if we are actually doing it: 'Mental representation, in the absence if direct visual or other stimulus – in other words, imaging – brings into play some of the same neurones that are involved in direct perception.' McGilchrist comments that, due to its close relationship with experience, 'We need to be careful of our imagination, since what we imagine is in a sense what we are and who we become.'[326] Huxley, who thought that for the man of letters 'outer reality is constantly related to the inner world of private experience', might have approved of these results:[327]

> The poet is, etymologically, the maker. Like all makers, he requires a stock of raw materials – in his case, experience. Now experience is not a matter of having actually swum the Hellespont, or danced with the dervishes, or slept in a doss-house. It is a matter of sensibility and intuition, of seeing and hearing the significant things, of paying attention at the right moments, of understanding and co-ordinating. Experience is not what happens to a man; it is what a man does with what happens to him.[328]

With our understanding of I-languages, Romain Rolland's similar words bring a different meaning to Huxley's 'experience':

> No one ever reads a book. He reads himself through books, either to discover or to control himself. And the most objective books are the most deceptive. The greatest book is not the one whose message engraves itself on the brain, as a telegraphic message engraves itself on the ticker-tape, but the one whose vital impact opens up other viewpoints, and from writer to reader spreads the fire that is fed by the various essences, until it becomes a vast conflagration leaping from forest to forest.[329]

Marion Milner's *On Not Being Able to Paint* echoes this, particularly in the following passage: 'Moments when the original poet in each of us created the outside world for us, by finding the familiar in the unfamiliar, are perhaps forgotten by most people; or else they are guarded in some secret place of memory because they were too much like visitations of the gods to be mixed with everyday thinking.'[330] For writers like McGilchrist, the left hemisphere's focus on precision and individuality has led to a sense that our visual systems are an alienating force to us. Johann Herder's major work on aesthetics and the visual arts seems to capture this tension: 'Sight destroys beautiful sculpture rather than creating it; it transforms it into planes and surfaces, and rarely does it not transform the beautiful fullness, depth and volume of sculpture into a mere play of mirrors.'[331] Similar feelings prompted Viktor Shklovsky to claim that 'art exists that one may recover the sensation of life; it exists to make one feel things, to make the stone *stony*. The purpose of art is to impart the sensation of things as they are perceived and not as they are known.'[332]

Louis Sass's extraordinary work demonstrated, among many other things, how 'when we cease to act, to be involved, spontaneous and intuitive, and instead become passive, disengaged, self-conscious, and stare in an "objective" fashion at the world around us, it becomes bizarre, alien, frightening.'[333] This notion will be of great importance to us later, but for now we should recognise that for Sass, 'as for Wittgenstein, there is a close relationship between philosophy and madness. The philosopher's "predilection for abstraction and alienation – for detachment from body, world and community" can produce a type of seeing and experiencing which is, in a literal sense, pathological.'[334] The act of staring, Wittgenstein noted, 'is closely bound up with the whole puzzle of solipsism.' Part of the reason for these feelings of detachment could be due to the fact that, as Kenneth Craik pointed out in 1943, 'Introspective psychology and analytic philosophy of the self, of perception and of will, do not seem to take into account that in any well made machine one is ignorant of the working of most of the parts – the better they work the less we are conscious of them ... It is only a fault that draws attention to the existence of a mechanism at all.'[335] Drew Leder qualifies that 'insofar as I perceive through an organ, it necessarily recedes from the perceptual field it discloses. I do not smell my nasal tissue, hear my ear, or taste my taste buds but perceive with and through such organs.'[336] The view Martin Foss takes both compliments Sass and Leder and is also suggestive of a remedy to the unfulfilling and exploitative nature of what anarchists and libertarian socialists have traditionally called 'wage slavery' (the remedy being to bring all aspects of one's life and work under one's control, and not supervised or dictated by an external authority):

> It is the most essential characteristic of the body that it *disappears as an independent thing* the more it fulfills its service, and that we aware of the only something is wrong, if some part does not serve, that is in sickness or tiredness.[337]

Antonio Damasio backs these suggestions with his neurology, writing in 1999 that 'We use part of the mind as a screen to prevent another part of it from sensing what goes on elsewhere.'[338] Consequently:

> The alleged vagueness, elusiveness, and intangibility of emotions and feelings is probably a symptom of this fact, an indication of how we cover the representation of our bodies, of how much mental imagery based on nonbody objects and events masks the reality of the body. Otherwise we would easily know that emotions and feelings are tangibly about the body. Sometimes we use our minds to hide a part of our beings from another part of our beings.
>
> I could describe the hiding of the body as a distraction, but I would have to add that it is a very adaptive distraction. In most circumstances, rather than concentrating resources on our inner states, it is perhaps more advantageous to concentrate one's resources on the images that describe problems out in the world or on the premises of those problems or on the options for their solution and their possible outcomes. Yet this skewing of perspective relative to what is available in our minds has a cost. It tends to prevent us from sensing the possible origin and nature of what we call self. When the veil if lifted, however, at the scale of understanding permitted to the human mind, I believe we can sense the origin of the construct we call self in the representation of individual life.[339]

Socrates, too, admitted that he 'was afraid that by observing objects with my eyes and trying to comprehend them with each of my other senses, I might blind my soul altogether.' It seems almost trivial, but there's clearly a reason why we don't introspect deeply when we're sliding down log flumes or playing basketball, as McGilchrist notes:

> Over-awareness itself alienates us from the world and leads to a belief that only we, or our thought processes, are real. If this seems curiously reminiscent of Descartes' finding that the only reliable truth was that his own thought processes guaranteed that he, at least, existed, that is not accidental. The detached, unmoving, unmoved observer feels that the world loses reality, becomes merely "things seen". Attention is focused on the field on consciousness itself, not on the world beyond, and we seem to experience *experience*. In his *Philosophical Investigations*, Wittgenstein actually notes that when this kind of staring attention takes over, others appear to lack consciousness, to be automata rather than minds (as Descartes had also found) ... There is a lack of seeing *through*, to whatever there is beyond.[340]

In his recent book *The Evolution of Childhood*, Melvin Konner has also detailed how between the ages of four and five the hormone andrenarche is secreted in the brain which aids directly in introspection.[341] Nicholas Humphrey's less helpful account of consciousness in *Soul Dust* makes the functionalist claim that to 'experience sensations "as having" these [phenomenal] features is to form a mental representation to that effect. ... Thus "consciousness" (or "being conscious"), as a state of mind, is the cognitive state of entertaining such mental representations.' Echoing Dennett, he thinks that consciousness 'is no more or less than a piece of magical "theater."' Before human consciousness rose to its current state, 'animals were engaged in some kind of inner monitoring of their own responses to sensory stimulation.'[342] But is calling consciousness simply an illusion solving the problem? Is it helpful do the same to free will? Is it even a naturalistic answer? Julian Jaynes wrote similar things in the 1970s. He argued (intriguingly, but unconvincingly) that consciousness is a 'virtual illusion,' that assumes the phenomenal properties of a theatre. When Humphrey writes that throughout evolutionary history our stimuli 'has become a virtual expression occurring at the level of a virtual body, hidden inside your head,' he is also repeating Thomas Nagel's idea that consciousness is simply 'what it is like to be' something. We should also be wary of Humphrey's invented terms, such as 'sensition' and 'ipsumdrum' – words which ensure the author's mark on the world but do little to further or explain naturalistic inquiry. In more recent times, the term 'qualia' has been used to refer to subjective conscious experience – three words which, in this case, are essentially synonymous. But as Samuel Guttenplan asks:

> Does the coining of the term "qualia" solve the problem with which we began? That is, can we now reconcile our knowledge of our experience with the demand that knowledge expressible? This certainly would be a quick fix to the original problem – almost certainly too quick. And any new word only counts as a genuine extension to our language if we are fairly sure we can understand what it is supposed to signify. And it is not clear that we have said enough to imbue "qualia" with a meaning suitable to get over the expressibility problem. For the suspicion is that this new word doesn't do any more for us than the demonstrative "that."[343]

David Deutsch, proponent of the Many-Worlds interpretation of quantum theory, also meditated on the virtual non-Platonic reality of mathematics in hearkening back to Hume's rejection of the 'theatre' of the mind:

Imagination is a straightforward form of virtual reality. What may not be so obvious is that our 'direct' experience of the world through our senses is virtual reality too. For our external experience is never direct; nor do we even experience the signals in our nerves directly – we would not know what to make of the streams of electrical crackles that they carry. What we experience directly is a virtual-reality rendering, conveniently generated for us by our unconscious minds from sensory data plus complex inborn and acquired theories (i.e. programs) about how to interpret them.

We realists take the view that reality is out there: objective, physical and independent of what we believe about it. But we never experience that reality directly. Every last scrap of our external experience is of virtual reality. And every last scrap of our knowledge – including our knowledge of the non-physical worlds of logic, mathematics and philosophy, and of imagination, fiction, art and fantasy – is encoded in the form of programs for the rendering of those worlds on our brain's own virtual-reality generator.[344]

This processing of experience through the adaption of memory and imagination 'becomes an instrumentality of success' in planning and action.[345] Both these traits 'provide a kind of isolation from the flux and press of sensory discrimination and response.' But with its 'freedom from representation in terms of sensory imagery, a language system is capable of greater independence from perceptual experience than is either memory or imagination.'[346] Though very little is known about the evolution of cognition, it should be uncontroversial that mental states, as part of an organism, have evolved and adapted. One might reasonably expect, with Dobzhansky, that 'nothing makes sense' in our mental experience except through an understanding of the brain's evolution. Perhaps surprisingly, Jung's psychoanalysis was finely tuned to this proposal: 'Just as the human body represents a whole museum of organs, with a long evolutionary history behind them, so we should expect the mind to be organized in a similar way ... We receive along with our body a highly differentiated brain which brings with it its entire history and when it becomes creative it creates out of this history – out of the history of mankind ... that age-old natural history which has been transmitted in living form since the remotest times, namely the history of the brain structure.'[347] Relating Russell's cognitive 'scopes and limits' to bodily restrictions, there is some evidence which suggests that 'the observer's perception of biological motion is guided and/or biased by his or her implicit knowledge of the biomechanical constraints that apply to the execution of his or her own movements.'[348] Jacob and Jeannerod ask us to

> Consider two pictures of a person, one with her left arm on the left side of her left knee and the other one with her left arm on the right side of her left knee. When such a pair of static images of human bodily postures is sequentially presented and the time interval between the two displays is within 550-750ms, then an observer sees the biomechanically possible apparent motion of the person's left arm move *around* her knee. When the same pair of static images in presented with a time interval of 150-350ms, an observer sees the biomechanically impossible apparent motion of the person's arm move *though* her knee.[349]

PET scans reveal that 'the primary motor cortex, the premotor and the inferior parietal cortex' are involved in the seeing of biomechanically possible apparent motion, whereas 'the perception of biomechanically impossible apparent motion prompted bilateral activation of the medial orbito-frontal cortex.'[350]

Recent discoveries about the nature of mirror neurons can also shed some light on the scopes of cognition, motor action and empathy. Having only been recently discovered at the University of Parma in Italy, mirror neurons help in the development of complex social interactions and are found in the frontal lobes. As their discoverers explain, when the monkey in their experiments 'observes a motor action that belongs (or resembles) its movement repertoire, this action is automatically retrieved. The retrieved action is not necessarily executed. It is only represented in the motor system. We speculate that this observation/execution mechanism plays a role in understanding the meaning of motor events.'[351] Certain motor-command neurons fire in the brain when I reach out my hand and pick up a grape in my desk. If I then move my arm to put the grape in my mouth, another set of motor-command neurons will fire. But Giacomo Rizzolatti and his colleagues discovered that a sub-set of these neurons (possibly twenty percent) will also fire when I watch someone else perform the same action – they 'are active both when we do something and when we watch others do it.'[352] Since their discovery, it's been largely suggested that mirror neurons must be involved in imitation and empathy.[353] Mirror neurons have also been interpreted as the basis of the 'theory of mind,' simulating the experience of others and helping us recognise, say, that when we see a man take an umbrella out of his rucksack it's because he thinks it's about to rain. V. S. Ramachandran also speculates

> that these neurons can not only help simulate other people's behavior but can be turned 'inward' – as it were – to create

> second-order representations or meta-representations of your own earlier brain processes. This could be the neural basis of introspection, and of the reciprocity of self awareness and other awareness. There is obviously a chicken-or-egg question here as to which evolved first, but ... The main point is that the two co-evolved, mutually enriching each other to create the mature representation of self that characterizes modern humans. ... [T]he only thing that separates you from me is a small subset of neural circuits in your frontal lobes interacting with mirror neurons.[354]

In an essay on self-awareness, he adds that

> self-awareness is simply using mirror neurons for "looking at myself *as if* someone else is looking at me" (the word "me" encompassing some my brain processes, as well). The mirror neuron mechanism – the same algorithm – that originally evolved to help you adopt *another's* point of view was turned inward to look at your own self.[355]

There are also mirrors neurons which recognise touch: if someone's arm is touched, a neuron in the sensory region of their brain fires, but the same neuron will also fire when that person watches someone else's arm being touched. But why do we not get confused and actually feel a sensation on our arm when we watch others being touched? This is because the touch and pain receptors in the skin send messages to the person's brain that his arm is not being touched; a sort of vetoing of the mirror neurons preventing the conscious experience of the touch sensation. But if his arm is then anesthetised, the person actually feels the touch sensation when he watches another person being touched, dissolving the barrier between each of them: 'What we feel arises out of what I feel for what you feel for what I feel about your feelings about me – and about many other things besides.'[356] This is due to the fact that certain 'single neurons, recorded mainly in area AIP of the dorsal pathway, respond preferentially to those geometrical properties of objects that serve such visuomotor tasks as grasping them.'[357] Montaigne would not have been surprised by these findings, since for him 'the very sight of someone else's pain causes me real pain, and my body often takes on the sensations of the person I am with. Another's perpetual cough tickles my lungs and throat. I'm more reluctant to visit those I love and am bound to care for, when they're sick, than those I care less about, and mean less to me. I adopt their disease that troubles me, and make it my own.'[358] In addition, psychologist Jean Decety and neuroscientist Thierry Chaminade concluded from their research in 2003 that 'The sense of self emerges from the activity of the brain in interaction with other selves.'[359] According to Max Scheler's proposals in *The Nature of Sympathy*, which have since been supported by research in child development since his death in 1928, 'our early experience of the world is intersubjective and does not include an awareness of self as distinct from other.'[360] There is 'an immediate flow of experiences, undifferentiated as between mine and thine, which actually contains both our own and other's experiences intermingled and without distinction from one another.'[361]

These elements of the developing 'mental' world are accurately characterised by Rodolfo Llinás' description of perception as 'a dream modulated by sensory input.' He adds that the mind is 'a computational state of the brain generated by the interaction between the external world and an internal set of reference frames.'[362] Dreaming is a mental state 'in which "cognition" does not relate to the co-existing reality.'[363] For him and Denis Paré, 'Dreaming and wakefulness are so similar from electrophysiological and neurological points of view that wakefulness may be described as a dreamlike state modulated by sensory input.'[364] Adding to this, 'the range of the brain's *unconscious* capacities is vastly extended in waking compared to sleep.'[365] It's also been shown that 'thinking using a concrete concept involves activating many of the same sensorimotor neural clusters that would be activated in actually perceiving something, manipulating an object, and moving one's body.'[366] Blurring the line further between Llinás' dreams and reality, the research of Marc Jeannerod shows that '*imagining* certain motor actions activates some of the same parts of the brain that are involved in actually performing that action. Imagining a visual scene also activates areas of the brain that would be activated if we actually perceived that scene.'[367] The neural clusters responsible for the visual processing of motion are also stimulated when subjects see pictures of actual and implied motion.[368]

Similar to Llinás, Plato described the process of understanding, yielded through capacities gained from previous lives (what today would be called genetic endowment), as the dawning of ideas which 'have been newly aroused in [us] as if in a dream,' from which an anti-foundationalism necessarily follows, since we can never ultimately justify our ideas, only use them.[369] Conversely, Descartes' anti-foundationalism was displayed in his *Principles of Philosophy*, where he 'explicitly rejects the mode used by ancient geometers proceeding deductively like Euclid, such as Pappus and Diophantus, as inadequately depicting (in fact as 'hiding' or 'covering up') how they actually discovered their truths.'[370] Perhaps helping us sympathise with countless mathematics teachers, the vindication of seventeenth-century anti-foundationalism demonstrates that the mind simply won't show us its working-out.

It's reasonable to suppose, as Russell did, that 'we cannot say whether' our perception of an object 'resembles the object in any intrinsic respect, except that both it and the object are brief events in space-time.'[371] Colour categories, for example, 'are determined jointly by the external physical world, human biology, the human mind, plus cultural considerations.'[372] As they stand, 'the brain sciences, despite important progress, are far from closing the gap by the problems posed by thought and language, or even to what is more or less understood about these topics.'[373] In terms of what they tell us about human nature as compared to other disciplines, David Brooks, writing in *The New Yorker*, believes that 'brain science helps fill the hole left by the atrophy of theology and philosophy.'[374]

Another distinction needs to be drawn between a person, his brain, thoughts, desires and so on. Namely, a *person* 'thinks, not their brains ... though their brains provide the mechanisms of thought. ... *People* in certain situations understand a language; my brain no more understands English than my feet take a walk.'[375] Intentional states are often ascribed to certain entities for the purposes of expressing and ordering human concerns. For example, whether a plane is flying or not is to be defined at the will of observer. The same is true of submarines that 'swim,' or robots and cats that 'think.' Intentional descriptions help us understand physics and animal behaviour, but our use of such descriptions seems to have no limits. We 'apply intentional description to all manner of phenomena that don't genuinely satisfy it: hurricanes, comets, flowers, sometimes the universe as a whole.'[376] This has had repercussions in various disciplines, and

> particularly troublesome in the twentieth century has been the degree to which it has become commonplace to ascribe all manner of often unconscious psychological states to people on the slenderest of objective bases – often merely because the ascription "makes sense" to, say, the patient or the therapist. The relatively inarticulate world of the small child seems to be a particularly inviting canvas for intentional description: consider the extraordinary thoughts attributed to young children by a Freud or a Melanie Klein, on the most amazingly meagre evidential base.[377]

It's consequently important to realise that ideas of 'unstoppable forces meeting immovable objects,' to take a popular example, are based on *linguistic* and not *physical* phenomena. Design and intention are human concepts, not universal laws. The ancient Japanese Zen master Huineng appears to have agreed with this assumption:

> Two monks were watching a flag flapping in the wind. One said to the other, "The flag is moving."
> The other replied, "The wind is moving."
> Huineng overheard this. He said, "Not the flag, not the wind; mind is moving."

If we were to discuss, for instance, 'the meaning of life,' despite the fact that such a notion has no technical definition, it does have a common sense one. It's quite like the story of the young Chesterton who was walking along the road with his mother and spied a man, asking 'Mummy, what's that man *for*?' 'Meaning' also applies to quite precise things. To take an example, I might initially find the sound of jazz music unpleasant and meaningless. After a few weeks of listening to it, though, I might come appreciate its rhythm and style, eventually finding it 'meaningful.' When we apply the same principle of 'meaning' to life itself, the sheer variety and enormity of things to experience overwhelms any general sense of meaning, and the concept becomes swollen out of definition. General questions of 'meaning' and 'purpose' are not directed at any specific aspect of the world, and so become incoherent. We find a suitable comparison to this in Wittgenstein: 'Mightn't we imagine a man who, never having had any acquaintance with music, comes to us and hears someone playing a reflective piece of Chopin and is convinced that this is a language and people merely want to keep the meaning from him?'[378] Even with this in mind, we may be satisfied with the ignorance of Sir Edward Appleton, who admitted: 'I do not mind what language an opera is sung in so long as it is a language I don't understand.'[379] As McGilchrist points out, it is primarily the task of the brain' left hemisphere ('the interpreter') to look for meaning in the world:

> Everything, just as it is, seems to have meaning, but what it is is never clear. The more one stares at things the more one freights them with import. That man crossing his legs, that woman wearing that blouse – it can't just be accidental. It has a particular meaning, is intended to convey something; but I am not let in on the secret, which every one else seems to understand. Notice that the focus of paranoia is a loss of the normal between-ness – something that should be conveyed from others to myself, is being kept from me. The world comes to appear threatening, disturbing, sinister.[380]

Before turning to a more specific, linguistic inquiry into the nature of meaning, Woody Allen's sardonic words perhaps express this particular issue more effectively:

> There's an old joke – um... two elderly women are at a Catskill mountain resort, and one of them says, "Boy, the food at this place is really terrible." The other one says, "Yeah, I know; and such small portions." Well, that's essentially how I feel about life – full of loneliness, and misery, and suffering, and unhappiness, and it's all over much too quickly.[381]

REFERENCES

[1] Noam Chomsky, 'Language from an Internalist Perspective,' *The Future of the Cognitive Revolution*, ed. David Martel Johnson and Christina E. Erneling (Oxford University Press, 1997), p. 132.
[2] Friedrich Albert Lange, *The History of Materialism and Criticism of its Present Importance*, trans. Ernest Chester Thomas, 3rd ed. (London: Routledge & Kegan Paul, 1957), p. 64.
[3] Andrew Goatly, *The Language of Metaphors* (London: Routledge, 1997), p. 3.
[4] Milan Kundera, *The Unbearable Lightness of Being*, trans. M. H. Heim (London: Harper Collins, 1991), p. 209.
[5] Cited in Raymond W. Gibbs, Jr., *The Poetics of Mind* (Cambridge University Press, 1994, repr. 1999), p. 181.
[6] Steven Pinker, *The Stuff of Thought: Language as a Window into Human Nature* (London: Penguin, 2008), 236.
[7] Goatly, *The Language of Metaphors*, p. 42.
[8] Pinker, *The Stuff of Thought*, p. 246.
[9] Richard Rorty, *Objectivity, Relativism and Truth* (Cambridge University Press, 1991), p. 129.
[10] Goatly, *The Language of Metaphors*, p. 84.
[11] Peter Stockwell, *Cognitive Poetics: An Introduction* (London: Routledge, 2002), p. 5.
[12] Pinker, *The Stuff of Thought*. p. 248.
[13] Iain McGilchrist, *The Master and His Emissary: The Divided Brain and the Making of the Western World* (Yale University Press, 2010), p. 170.
[14] Gibbs, *The Poetics of Mind*, p. 5.
[15] V. S. Ramachandran and E. M. Hubbard, 'Synaesthesia: a window into perception, thoughts and language,' *Journal of Consciousness Studies*, 8(12), 2001: 3-34.
[16] http://mind.cog.jhu.edu/faculty/landau/AllWebsitePublications/2010LandauDessalegnGoldberg.pdf
[17] C. R. Gallistel, 'The Foundational Abstractions,' *Of Minds and Language: A Dialogue with Noam Chomsky in the Basque Country*, ed. Massimo Piatello-Palmarini, Juan Uriagereka and Pello Salaburu (Oxford University Press, 2009), p. 63 (58-73).
[18] Ibid., p. 64.
[19] Catherine de Lange, 'My two minds,' *New Scientist*, 5 May 2012.

[20] Scott P. Johnson, 'How Infants Learn About the Visual World,' *Cognitive Science*, 34(7), September 2010: 1158-1184.
[21] Gibbs, *The Poetics of Mind*, p. 87 (emphasis his).
[22] Walt Whitman, 'Slang in America,' in F. Stovall ed., *The Collected Writings of Walt Whitman: Prose Works* (New York University Press, 1964), p. 573.
[23] Richard Rorty, *Philosophy and the Mirror of Nature* (Princeton University Press, 1989), p. 18.
[24] Richard Lederer, *Crazy English* (Simon and Schuster, 1998), p. 15.
[25] Ibid., p. 11 (emphasis his).
[26] Franz Boas, *Handbook of American Indian Languages* (1973), cited in John A. Lucy, *Language Diversity and Thought: A Reformulation of the Linguistic Relativity Hypothesis* (Cambridge University Press, 1996 [1992]), p. 15. For an interesting discussion and rebuttal of the linguistic relativity argument hinted at, see Noam Chomsky's introduction to Adam Schaff, *Language and Cognition* (New York: McGraw Hill, 1973).
[27] Leonardo da Vinci, *Notebooks*, ed. Thereza Wells (Oxford University Press, 2008), p. 3.
[28] L. H. Finkel, 'Neuronal group selection: a basis for categorisation by the nervous system,' in D. de Kerckhove and C. J. Lumsden (eds.), *The Alphabet and the Brain: The Lateralization of Writing* (Berlin: Springer-Verlag, 1988), pp. 51-70.
[29] Cited in Galen Strawson, 'Real Materialism,' *Chomsky and His Critics*, eds. Louise M. Antony and Norbert Hornstein (Oxford: Blackwell, 2003), p. 63.
[30] Cited in Michael Brooks, 'Weirdest of the weird,' *New Scientist*, 8 May 2010.
[31] Lera Boroditsky, 'How Language Shapes Thought,' *Scientific American*, February 2011.
[32] Joseph Priestley, *Priestley's Writings on Philosophy, Science, and Politics*, ed. John A. Passmore (New York: Collier Books, 1965).
[33] Alan Turing, 'Computing Machinery and Intelligence,' in *The Essential Turing*, ed. B. Jack Copeland (Oxford: Clarendon Press, 2004), p. 449. Originally published in *Mind*, (1950) 49: 433-60.
[34] John Duncan, *How Intelligence Happens* (Yale University Press, 2010), p. 26.
[35] Ray Jackendoff, *Foundations of Language: Brain, Meaning, Grammar, Evolution* (Oxford University Press, 2002), p. 108, n. 1.
[36] David Hume, *An Inquiry Concerning Human Understanding and Concerning the Principles of Morals* (Oxford: Clarendon Press, 1988), p. 13.
[37] Alan Turing, 'Can Digital Computers Think?', a lecture first broadcast on BBC Radio on 15 May 1951, in *The Essential Turing*, p. 486.
[38] 'Can Automatic Calculating Machines Be Said To Think?', a discussion between Turing, R. B. Braithwaite, Geoffrey Jefferson, and Max Newman broadcast on BBC Radio on 14 January 1952, ibid., p. 494.
[39] Alan Turing, 'Intelligent Machinery,' ibid., p. 431.
[40] Ibid., p. 410.
[41] Turing, 'Can Automatic Calculating Machines Be Said To Think?', ibid., pp. 494-5.
[42] Turing, 'Intelligent Machinery,' ibid., p. 431.
[43] Ray Jackendoff, *Language, Consciousness, Culture: Essays on Mental Structure* (Massachusetts: MIT Press, 2007), pp. 104-5.
[44] Albert Einstein, 'A letter to Jacques Hadamard,' cited in McGilchrist, *The Master and His Emissary*, p. 107.
[45] David Robson, 'The voice of reason,' *New Scientist*, 4 September 2010.
[46] Ibid. See Jeffrey Loewenstein and Dedre Gentner, 'Relational language and the development of relational mapping,' *Cognitive Psychology*, 50(4): 315-353.
[47] Gary Lupyan, 'The Conceptual Grouping Effect: Categories Matter (and named categories matter more),' *Cognition*, 108, 2008: 566-577.

[48] However, see Alison Motluk, 'It's not what you've got...', *New Scientist*, 31 July 2010, for a review of recent evidence which suggests human intelligence 'can no longer be attributed to a "supersized" brain.'
[49] R. Varley and M. Siegal, 'Evidence for cognition without grammar from causal reasoning and 'theory of mind' in an agrammatic aphasic patient,' *Current Biology*, 10(12): 723-6.
[50] Richard Elwes, 'It doesn't add up,' *New Scientist*, 14 August 2010. See Elwes' book *Maths 1001: Absolutely Everything that Matters in Mathematics* (Quercus, 2010).
[51] McGilchrist, *The Master and His Emissary*, p. 65.
[52] Ibid., p. 109.
[53] Neil Smith, 'Chomsky's science of language,' *The Cambridge Companion to Chomsky*, p. 41.
[54] Noam Chomsky, introduction to Adam Schaff, *Language and Cognition*, ed. Robert S. Cohen, trans. Olgierd Wojtasiewicz (New York: McGraw Hill, 1973), p. v.
[55] McGilvray, *The Cambridge Companion to Chomsky*, p. 308, n. 5.
[56] Noam Chomsky, *New Horizons in the Study of Language and Mind* (Cambridge University Press, 2000), p. xii.
[57] Pierre Jacob, 'Chomsky, Cognitive Science, Naturalism and Internalism,' 2001, http://hal.inria.fr/docs/00/05/32/33/PDF/ijn_00000027_00.pdf, p. 52.
[58] Murray Gell-Mann, in Matthew Blakeslee, *New Scientist*, 24 July 2010.
[59] Walter J. Gehring and Kazuko Ikeo, *Trends in Genetics*, 15(9), September 1999: 371-377.
[60] Chomsky, *New Horizons in the Study of Language and Mind*, p. 104.
[61] *Thinking Allowed*, 'Conversations on the leading edge of knowledge and discovery with Dr. Jeffrey Mishlove,' www.williamjames.com/transcripts/pinker1.htm.
[62] Cited in Cory Juhl and Eric Loomis, *Analyticity* (London: Routledge, 2010), p. 62.
[63] Colin McGinn, 'Consciousness and Space,' www.nyu.edu/gsas/dept/philo/courses/consciousness97/papers/ConsciousnessSpace.html.
[64] John Collins, 'Naturalism in the Philosophy of Language; or Why There Is No Such Thing as Language,' *New Waves in Philosophy of Language*, ed. Sarah Sawyer (London: Palgrave Macmillan, 2010), p. 56, n. 2 (41-59).
[65] Leo Tolstoy, *War and Peace*, trans. Louise and Aylmer Maude, ed. Henry Gifford (Oxford University Press, 1991, repr. 2008), p. 1304 (emphasis mine).
[66] Yevgeny Zamyatin, *We*, trans. Mirra Ginsburg, http://leecworkshops.wikispaces.com/file/view/yevgeny-zamyatin-we.pdf.
[67] Rafael Núñez, 'Conceptual Metaphor, Human Cognition, and the Nature of Mathematics,' *The Cambridge Handbook of Metaphor and Thought*, ed. Raymond W. Gibbs, Jr. (Cambridge University Press, 2008), p. 343.
[68] Ibid., p. 358.
[69] Saunders MacLance, 'Mathematical Models: A Sketch for the Philosophy of Mathematics,' *American Mathematical Monthly*, August-September 1981, p. 470.
[70] Aristotle, *Metaphysics*, M3, 1077b 17-20, cited in Gordon C. Brittan Jr., 'Towards a Theory of Theoretical Objects,' *PSA: Proceedings of the Biennial Meeting of the Philosophy of Science Association*, 1986(1): 388 (384-393).
[71] Werner Heisenberg, cited in Subrahmanyan Chandrasekhar, *Truth and Beauty: Aesthetics and Motivations in Science* (University of Chicago Press, 1987), p. 65.
[72] Werner Heisenberg, *Physics and Philosophy* (London: Allen & Unwin, 1963), p. 177.
[73] Michael Brooks, *New Scientist*, 23 July 2011.
[74] Vlatko Vedral, 'Living in a Quantum World,' *Scientific American*, June 2011.
[75] Cited in Gordon C. Brittan Jr., 'Towards a Theory of Theoretical Objects,' *PSA: Proceedings of the Biennial Meeting of the Philosophy of Science Association*, 1986(1): 389 (384-393).

[76] Aldous Huxley, *The Doors of Perception and Heaven and Hell* (London: Flamingo, 1994 [1954]), p. 10.
[77] Brian Greene, interview with Amanda Gefter, 'Thoughts racing along parallel lines,' pp. 30-31, *New Scientist*, 5 February 2011.
[78] Galen Strawson, 'The Self,' *The Oxford Handbook of Philosophy of Mind*, ed. Brian P. McLaughlin, Ansgar Beckermann and Sven Walter (Oxford University Press, 2009), p. 541.
[79] Cited in Louis Scutenaire, *René Magritte* (Brussels: Librarie Sélection, 1947), p. 83.
[80] Amanda Gefter, 'Countdown to the theory of everything,' *New Scientist*, 8 October 2011.
[81] Linda Geddes, 'The clock in your head,' *New Scientist*, 8 October 2011.
[82] Cited in Stuart Clark, 'The origin of time,' *New Scientist*, 8 October 2011.
[83] Ibid.
[84] Gefter, *New Scientist*, 8 October 2011. See *New Scientist*, 19 January 2008.
[85] Ibid.
[86] Descartes, 'Meditation VI' in *The Philosophical Writings of Descartes*, trans. J. Cottingham, R. Stoothoff, D. Murdoch and A. Kenny (Cambridge University Press, 1985/1618-28) vol. II, p. 54.
[87] Saul Kripke, *Naming and Necessity* (Harvard University Press, 1980), p. 355, n. 77.
[88] Jaegwon Kim, 'The mind-body problem,' in *The Oxford Companion to Philosophy*, ed. T. Honderich (Oxford University Press, 1995), p. 579.
[89] Cited in *The Philosophical Writings of Descartes*, vol. III, p. 220.
[90] Jaegwon Kim, 'Mental causation,' *The Oxford Handbook of Philosophy of Mind*, p. 31.
[91] Chomsky, *New Horizons in the Study of Language and Mind*, p. 109.
[92] Noam Chomsky, *On Nature and Language* (Cambridge University Press, 2002), p. 68.
[93] Stephen Gaukroger, *The Emergence of a Scientific Culture: Science and the Shaping of Modernity, 1210-1685* (Oxford University Press, 2006), p. 16.
[94] Geoffrey Hellman, 'Determination and Logical Truth,' *Journal of Philosophy*, 82, 1985: 609.
[95] Cited in Richard C. Allen, *David Hartley on Human Nature* (State University of New York Press, 1999), p. 377.
[96] Edmund Burke, *A Philosophy Enquiry into the Origin of our Ideas of the Sublime and Beautiful*, ed. Adam Phillips (Oxford University Press, 1990/1757), p. 12.
[97] Chomsky, *New Horizons in the Study of Language and Mind*, p. 106. See also Joachim Schummer, 'Towards a Philosophy of Chemistry,' *Journal for General Philosophy of Science*, 28(2), 1997: 307-36.
[98] Gerald Horton, *Striking Gold in Science: Fermi's Group and the Recapture of Italy's Place in Physics* (London: Minerva, 1974), p. 165.
[99] Jackendoff, *Language, Consciousness, Culture*, p. 13.
[100] Ibid., p. 15.
[101] John Collins, *New Waves in Philosophy of Language*, p. 44.
[102] Jackendoff, *Language, Consciousness, Culture*, p. 4.
[103] Gilbert Harman, Review of New Horizons in the Study of Language and Mind by Noam Chomsky, *The Journal of Philosophy* 98(5), May 2001: 266 (265-9).
[104] Lange, *The History of Materialism and Criticism of its Present Importance*, p. 138.
[105] Cited in ibid., p. 17.
[106] Max Born, 1954 Nobel Lecture, 'The statistical interpretation of quantum mechanics,' cited in Paul Tappenden, 'Evidence and Uncertainty in Everett's Multiverse,' *The British Journal for the Philosophy of Science*, 62(1), March 2011: 99-123. Text available online: nobelprize.org/nobel_prizes/physics/laureates/1954/born-lecture.pdf.
[107] Patricia Churchland, 'Presidential address of the APA Pacific Division,' March 1994.
[108] Chomsky, *New Horizons in the Study of Language and Mind*, p. 113.

[109] Wolfram Hinzen, *Mind Design and Minimal Syntax* (Oxford University Press, 2006), pp. 56-7.
[110] John Searle**Error! Bookmark not defined.**, *The Rediscovery of the Mind* (MIT Press, 1992), p. 103.
[111] Cited in Mark Buchanan, 'Why nature is not the sum of its parts,' *New Scientist*, 4 October 2008.
[112] Noam Chomsky, *The Science of Language: Interviews with James McGilvray* (Cambridge University Press, 2012), p. 74.
[113] Andrew Melnyk, *A Physicalist Manifesto: Thoroughly Modern Materialism* (Cambridge University Press, 2003), p. 2 (emphasis his).
[114] Chris Daly, 'What Are Physical Properties?', *Pacific Philosophical Quarterly*, 79, 1998: 213.
[115] Pinker, *The Stuff of Thought*, p. 224.
[116] Ibid.
[117] Lange, *The History of Materialism and Criticism of its Present Importance*, p. 92.
[118] Russell, 'Why I am not a Christian,' *The Basic Writings of Bertrand Russell, 1903-1959*, ed. Robert E. Egner and Lester E. Dennon (London: Routledge, 1999), p. 586.
[119] Lange, *The History of Materialism and Criticism of its Present Importance*, p. 99.
[120] Ibid., p. 101.
[121] Bertrand Russell, 'On the Notion of Cause,' *Mysticism and Logic* (New York: Doubleday, Anchor Books, 1957), p. 174.
[122] A. N. Whitehead, *The Concept of Nature* (Cambridge University Press, 1920), p. 57, 73.
[123] Anthony Flew (ed.), *A Dictionary of Philosophy* (London: Macmillan, 1979), p. 74.
[124] Lange, *The History of Materialism and Criticism of its Present Importance*, pp. 173-4.
[125] Michael Brooks and Richard Webb, 'All of nothing,' *New Scientist*, 12 May 2012.
[126] John Ziman, *Public Knowledge* (Cambridge University Press, 1968), p. 9.
[127] Cited in Hinzen, *Mind Design and Minimal Syntax*, p. 4.
[128] Ibid.
[129] Derek Fraser, *The Evolution of the British Welfare State*, 3rd ed. (London: Macmillan, 2002).
[130] Steven Weinberg, 'The Forces of Nature,' *Bulletin of the American Academy of Arts of Sciences*, 29(4), 1976: 13-29.
[131] Cited in Hinzen, *Mind Design and Minimal Syntax*, p. 6.
[132] Ibid., p. 54.
[133] Plato, *Statesman*, trans. and ed. C. J. Rowe (Warminster: Aris and Phillips, 1995), 277d1-4.
[134] Huxley, *The Doors of Perception and Heaven and Hell*, p. 52.
[135] Hinzen, *Mind Design and Minimal Syntax*, p. 59.
[136] Priestley, 'Of the Cause of Volition and the Nature of the Will,' *Priestley's Writings on Philosophy, Science, and Politics*, p. 65.
[137] Michael S. Gazzaniga, Richard B. Ivry and George R. Mangun with Megan S. Steven, *Cognitive Neuroscience: The Biology of the Mind*, 3rd ed (London: W. W. Norton & Company, 2009), p. 4.
[138] G. P. Thomson, *J. J. Thomson and the Cavendish Laboratory in his Day* (London: Nelson, 1964), p. 114.
[139] Cited in Derek Gjersten, *Science and Philosophy: Past and Present* (Oxford University Press, 1989), p. 20.
[140] Ibid., p. 21, 22.
[141] Cited in Hannah Arendt, *The Life of the Mind* (London: Harcourt, 1977), pp. 7-8.
[142] Marjorie McCorquodale, 'Poets and Scientists,' *Bulletin of the Atomic Scientists*, November 1965, p. 20.
[143] Aldous Huxley, *Literature and Science* (London: Chatto & Windus, 1963), p. 67.

[144] John D. Barrow, 'In the world's image,' *Times Higher Educational Supplement*, 16 April 1993.
[145] McGilchrist, *The Master and His Emissary*, p. 385.
[146] Paisley Livingston, 'Poincaré's Delicate Sieve: On Creativity and Constraints in the Arts,' in Michael Krausz, Denis Dutton and Karen Bardsley (eds.), *Philosophy of History and Culture, Volume 28: The Idea of Creativity* (Boston: Brill Academic Publishers), p. 130.
[147] Edward O. Wilson, *The Diversity of Life* (Harvard University Press, 1992), p. 5.
[148] Wilhelm von Humboldt, *The Limits of State Action*, ed. J. W. Burrow (Cambridge University Press, 1969).
[149] Huxley, *Literature and Science*, p. 91.
[150] Nancy Cartwright, Jordi Cat, Lola Fleck and Thomas E. Uebel, *Otto Neurath: Philosophy between Science and Politics* (Cambridge University Press, 1996), p. 253.
[151] David Knight, *Humphry Davy: Science & Power* (Cambridge University Press, 1998), p. 87.
[152] Cited in Richard Dawkins, *The God Delusion* (London: Bantam Press, Black Swan edition, 2007), p. 70-71.
[153] Arthur Eddington, *The Nature of the Physical World* (New York: Macmillan, 1928), pp. 51-2.
[154] Cited in Christopher Hill, *The World Turned Upside Down: Radical Ideas during the English Revolution* (London: Temple Smith, 1972), p. 71.
[155] Joseph Campbell, *Myths to Live By* (New York: Bantam, 1972), p. 10.
[156] Huxley, *Literature and Science*, p. 70 (emphasis his).
[157] Strawson, *Real Materialism and Other Essays*, p. 197, n. 23.
[158] Pynchon, *Gravity's Rainbow*.
[159] Lange, *The History of Materialism and Criticism of its Present Importance*, p. 155.
[160] Ibid.
[161] David Hume, *Enquiries Concerning Human Understanding and Concerning the Principles of Morals*, ed. L. A. Selby-Bigge (Oxford: Clarendon Press, 1975/1748), p. 63.
[162] *A Treatise on Human Nature*, Book I, Part 3, Section 2.
[163] Bertrand Russell, 'Introduction: Materialism, Past and Present,' Lange, *The History of Materialism and Criticism of its Present Importance*, p. v.
[164] David Papineau, 'The causal closure of the physical and naturalism,' *The Oxford Handbook of Philosophy of Mind*, pp. 53-4.
[165] A. Leslie, 'The necessity of illusion: Perception and thought in infancy,' *Thought without Language*, ed. L. Weiskrantz (Oxford University Press, 1988), pp. 185-210; E. S. Spelke, 'The origins of physical knowledge,' *Thought without Language*, ed. L. Weiskrantz (Oxford University Press, 1988).
[166] D. S. Berry and K. Springer, 'Structure, motion, and preschoolers' perception of social causality,' *Ecological Psychology*, 5, 1993: 273-83.
[167] F. Heider and M. Simmel, 'An experimental study of apparent behavior,' *American Journal of Psychology*, 57, 1944: 243-59.
[168] Parmenides, 'Poem of Parmenides: On nature,' *Fragments*, trans. and ed. A. Fairbanks, in A. Fairbanks, *The First Philosophers of Greece* (London: Kegan Paul, Trench, Trubner & Co., 1898), fr. DK B3.
[169] Stephen Stich, *From Folk Psychology to Cognitive Science* (MIT Press, 1983), p. 14.
[170] Steven Pinker, *The Blank Slate: The Modern Denial of Human Nature* (London: Penguin, 2002), p. 70. See Hilary Putnam, 'Reductionism and the nature of psychology,' *Cognition*, 2, 1973: 131-46.
[171] Matthew Blakeslee, 'I loved it. I had a knack,' interview with Murray Gell-Mann, *New Scientist*, 24 July 2010.
[172] Russell, 'Why I Took to Philosophy,' *The Basic Writings of Bertrand Russell*, p. 56.

[173] Joseph Priestley, *Disquisitions Relating to Matter and Spirit* (Kessinger Publishing, 1777), p. 29.
[174] Priestley, *Priestley's Writings on Philosophy, Science, and Politics*, p. 113, 115.
[175] David Lightfoot, 'Natural Selection-Itis,' *The Oxford handbook of Language Evolution*, ed. Maggie Tallerman and Kathleen R. Gibson (Oxford University Press, 2012), p. 313 (pp. 313-7).
[176] Ibid., p. 316.
[177] Hinzen, *Mind Design and Minimal Syntax*, p. 61.
[178] Marx X. Wartofsky, *Conceptual Foundations of Scientific Thought: An Introduction to the Philosophy of Science* (New York: Macmillan, 1968), p. 370.
[179] Cited in Noam Chomsky, 'The Mysteries of Nature: How Deeply Hidden?', *Chomsky Notebook*, ed. Jean Bricmont and Julie Franck (Columbia University Press, 2010), p. 5.
[180] John Dewey, *Human Nature and Conduct* (New York: Henry Holt and Company, 1922), p. 62.
[181] Cited in Chomsky, 'The Mysteries of Nature: How Deeply Hidden?', *Chomsky Notebook*, p. 5.
[182] Francesco Algarotti, *Sir Isaac Newton's Philosophy Explain'd For the Use of the Ladies*, trans. Elizabeth Carter (London: E. Cave, 1739/1737[first published]), vol. 2, p. 194.
[183] Galen Strawson, 'Real naturalism,' Lecture at the Human Experience and Nature, Royal Institute of Philosophy Conference 2011, University of the West of England, Bristol: http://www.youtube.com/watch?v=30laetzUf2Y.
[184] Eddington, *The Nature of the Physical World*, p. 291.
[185] McGilchrist, *The Master and His Emissary*, p. 7.
[186] Jeffrey Poland, 'Chomsky's Challenge to Physicalism,' in *Chomsky and His Critics*, p. 39.
[187] Wartofsky, *Conceptual Foundations of Scientific Thought*, p. 351.
[188] David Hume, *A Treatise of Human Nature*, ed. L. A. Selby-Bigge (Oxford: Clarendon, 1958 [1888]) Book I, Part IV, Section VI, 'Of personal identity,' p. 253.
[189] Russell, *The Basic Writings of Bertrand Russell*, p. 32.
[190] G. Marcus, *The Algebraic Mind: Integrating Connectionism and Cognitive Science* (MIT Press, 2001).
[191] D. H. Lawrence, *Study of Thomas Hardy and other essays*, ed. Bruce Steele (Cambridge University Press, 1985), p. 193.
[192] Cited in M. Capek, *The Philosophical Impact of Contemporary Physics* (Princeton, New Jersey: D. Van Nostrand, 1961) p. 319.
[193] Giovanni Vignale, 'The power of the abstract,' *New Scientist*, 26 February 2011.
[194] E. J. Lowe, *Routledge Philosophy Guidebook to Locke on Human Understanding* (Routledge, T. J. Press Ltd, Cornwall, 1995), p. 55 (emphasis his).
[195] Allen, *David Hartley on Human Nature*, p. 93.
[196] Chomsky, *New Horizons in the Study of Language and Mind*, p. 144-45.
[197] Huxley, *Literature and Science*, p. 11.
[198] Margaret Jacob, *Living the Enlightenment: Freemasonry and Politics in Eighteenth-Century Europe* (Oxford University Press, 1991), p. 200.
[199] Descartes, 'Principles of Philosophy,' Part Two, Principle 37, *The Philosophical Writings of Descartes*, trans. J. Cottingham, R. Stoothoff, D. Murdoch and A. Kenny (Cambridge University Press, 1985/1618-28) vol. I, pp. 240-41.
[200] Noam Chomsky, *Language and Problems of Knowledge: The Managua Lectures*, (MIT Press, 1987), p. 142.
[201] Chomsky, in Guttenplan, *A Companion to the Philosophy of Mind*, p. 157.
[202] Michael Friedman, 'Remarks on the history of science and the history of philosophy,' in *World Changes: Thomas Kuhn and the Nature of Science*, ed. P. Howich (MIT Press, 1993), p. 48 (emphasis his).

[203] Galen Strawson, 'Real Materialism,' *Chomsky and His Critics*, p. 65.
[204] William G. Lycan, 'Chomsky on the Mind-Body Problem,' *Chomsky and His Critics*, p. 25.
[205] Bertrand Russell, *An Outline of Philosophy* (London: Routledge, 1992 [1927]), p. 125.
[206] Vignale, *New Scientist*, 26 February 2011.
[207] Bertrand Russell, *Mysticism and Logic* (*New York*: Longmans, Green and Co., 1918), ch. 4.
[208] Steven Weinberg, 'The Forces of Nature,' *Bulletin of the American Academy of Arts of Sciences*, 29(4), 1976: 13-29.
[209] Jacob, 'Chomsky, Cognitive Science, Naturalism and Internalism,' 2001, http://hal.inria.fr/docs/00/05/32/33/PDF/ijn_00000027_00.pdf, pp. 49-50.
[210] Stephen Weinberg, 'The Forces of Nature,' *Bulletin of the American Academy of Arts and Sciences* (January 1976), 29 (4): 29-29.
[211] Russell, *The Problems of Philosophy*, p. 19.
[212] Chomsky, *New Horizons in the Study of Language and Mind*, pp. 89-90.
[213] Cited in Chomsky, 'The Mysteries of Nature: How Deeply Hidden?', *Chomsky Notebook*, p. 1.
[214] Letter to Bentley, 1693, cited in Chomsky, 'The Mysteries of Nature: How Deeply Hidden?' *Chomsky Notebook*, p. 6.
[215] Cited in *John Locke: Critical Assessments*, ed. Richard Ashcraft, vol. IV (London: Routledge, 1991), p. 19.
[216] John Locke, *An Essay Concerning Human Understanding*, ed. A. D. Woozley (Glasgow: William Collins Sons & Co Ltd, 1964, repr. 1984), p. 164.
[217] Richard Rorty, 'Method, Social Science, and Social Hope,' *Consequences of Pragmatism* (University of Minnesota, 1986), cited in, *Rorty and His Critics*, ed. Robert B. Brandom (Oxford: Blackwell, 2001), p. xiii.
[218] Gerald Feinberg, 'Ordinary Matter,' *Scientific American*, May 1967.
[219] Lange, *The History of Materialism and Criticism of its Present Importance*, p. 308.
[220] Huxley, *Literature and Science*, p. 10.
[221] Xenophon, *Conversations of Socrates*, ed. Robin Waterfield, trans. Hugh Tredennick and Robert Waterfield (London: Penguin, 1990), p. 71.
[222] Ludwig Wittgenstein, *Philosophical Investigations*, trans. G. E. M. Anscombe (London: Blackwell, 1997), p. 232.
[223] Xenophon, *Conversations of Socrates*, p. 70.
[224] Russell, *The Basic Writings of Bertrand Russell*, Preface, p. 7.
[225] Cited in Richard Dawkins, *The Oxford Book of Modern Science Writing* (Oxford University Press, 2009), p. 12.
[226] Strawson, 'Real Materialism,' *Chomsky and His Critics*, p. 64.
[227] Thomas Samuel Kuhn, *The Copernican Revolution: Planetary Astronomy in the Development of Western Thought* (Harvard University Press, 2003), p. 259.
[228] *Scientific American*, July 2011.
[229] John Duncan, 'Inside the puzzle of smart brains,' *New Scientist*, 13 November 2010.
[230] David Robson, 'A brief history of the brain,' *New Scientist*, 24 September 2011.
[231] Daniel Dennett, *Consciousness Explained* (London: Penguin, 1993), p. 187.
[232] Freeman Dyson, *Disturbing the Universe* (New York: Harper & Row, 1979), p. 50.
[233] Amanda Gefter, 'Why is there something rather than nothing?', *New Scientist*, 23 July 2011.
[234] Ibid.
[235] Ibid.
[236] Ibid.
[237] Christiaan Huygens, *Cosmotheoros* (London: Timothy Childe, 1698), p. 160.
[238] Cited in Allen, *David Hartley on Human Nature*, p. 96.
[239] Ibid., p. 97.

[240] Arthur Schopenhauer, *The Wisdom of Schopenhauer*, trans. Walter Jekyll (London: Watts & Co., 1911), pp. 65-6.
[241] Pierre Jacob, 'The Scope and Limits of Chomsky's Naturalism,' *Chomsky Notebook*, p. 230.
[242] V. S. Ramachandran, *The Tell-Tale Brain: A Neuroscientist's Quest for What Makes Us Human* (W. W. Norton & Company, 2011).
[243] Douglas Fox, 'The Greatest Map of All,' *New Scientist*, 5 February 2011.
[244] Ramachandran, *The Tell-Tale Brain*.
[245] Gatsby Computational Neuroscience Unit, UCL, www.gatsby.ucl.ac.uk/research.html.
[246] Gerald M. Edelman, *Bright Air, Brilliant Fire: On the Matter of the Mind* (New York: Basic Books, 1993).
[247] John Duncan, *New Scientist*, 13 November 2010.
[248] Jaegwon Kim, *Philosophy of Mind* (Westview Press Inc., 1998), p. 101.
[249] C. R. Gallistel, 'Neurons and Memory,' in M. S. Gazzaniga (ed.), *Conversations in the Cognitive Neurosciences* (MIT Press, 1997).
[250] Chomsky, *Problems of Knowledge and Freedom*, p. 31.
[251] David Chalmers, in Susan Blackmore, *Conversations on Consciousness* (Oxford University Press, 2005), p. 42.
[252] Marc Hauser's response to V. S. Ramachandran's essay, 'Self Awareness: The Last Frontier,' *Edge*, 1 January 2009, www.edge.org/3rd_culture/rama08/rama08_index.html.
[253] Timothy D. Wilson and Elizabeth W. Dunn, 2004,'Self-Knowledge: Its Limits, Value, and Potential for Improvement,' *Annual Review of Psychology* 55 (1), p. 507.
[254] Galen Strawson, 'Realm Materialism,' *Chomsky and His Critics*, p. 64.
[255] Immanuel Kant, *Critique of Pure Reason*, trans. Norman Kemp Smith (London: St Martin's Press, 1963), p. 166.
[256] Werner Heisenberg, in Marjorie McCorquodale, 'Poets and Scientists,' *Bulletin of the Atomic Scientists*, November 1965, p. 20.
[257] Bertrand Russell, *An Inquiry into Meaning and Truth: The William James Lectures for 1940 Delivered at Harvard University* (London: Routledge, 1995), Introduction.
[258] Rodolfo Llinás, ''Mindness' as a functional state of the brain,' *Mindwaves: Thoughts on Intelligence, Identity and Consciousness*, ed. Colin Blakemore and Susan Greenfield (Oxford: Blackwell, 1987), pp. 351-2 (339-58).
[259] Brian O'Shaughnessy, 'Observation and the Will,' *Journal of Philosophy*, 1963, cited in Guttenplan, *A Companion to the Philosophy of Mind*, p. 13.
[260] Ibid., p. 14.
[261] 'The Ideas of Chomsky,' interview with Bryan Magee, BBC, www.youtube.com/watch?v=EksuA4IAQIk.
[262] Milena Nuti, *Ethnoscience: Examining Common Sense*, PhD thesis, University College London (2005), p. 7.
[263] Chomsky, 'The Mysteries of Nature: How Deeply Hidden?' *Chomsky Notebook*, p. 11.
[264] Thomas Huxley, *Science and Education*, vol. III of *Collected Essays* (New York: Greenwood Press, 1968), pp. 39-40.
[265] Voltaire, *Candide: Or, Optimism*, trans. and ed. Theo Cuffe (London: Penguin Classics Deluxe Edition, 2005), p. 106.
[266] Schopenhauer, *The Wisdom of Schopenhauer*, p. 20.
[267] Cited in Noam Chomsky, *Chomsky on Anarchism*, ed. Barry Pateman (Edinburgh: AK Press, 2009), p. 104.
[268] Georges Rey, 'Chomsky, Intentionality, and a CRTT,' *Chomsky and His Critics*, p. 112.
[269] Donald Davidson, 'The Perils and Pleasures of Interpretation,' *The Oxford Handbook of Philosophy of Language*, ed. Ernest Lepore and Barry C. Smith (Oxford: Clarendon, 2006), p. 1056.
[270] Ibid., p. 1057.

[271] William James, *Principles of Psychology*, 1890, vol. I, cited in Wayne Wu, 'What is Conscious Attention?', pp. 93-120, *Philosophy and Phenomenological Research*, LXXXII(1), January 2011: 93.
[272] Cited in Jack J. Pasternak, *An Introduction to Human Molecular Genetics: Mechanisms of Inherited Diseases* (New Jersey: John Wiley & Sons, 2nd ed., 2005), p. 439.
[273] N. Kanwisher and D. Downing, 'Separating the Wheat from the Chaff,' *Science*, 282(5386), 1998: 57-8.
[274] Semir Zeki, 'Art and the Brain,' *Dædalus*, 127(2), Spring 1998.
[275] Cited in Leonard Shlain, *Art & Physics: Parallel Visions in Space, Time and Light* (New York: HarperCollins Perennial, 2001), p. 33.
[276] Jessica Griggs, 'Why your brain flips over visual illusions,' *New Scientist*, 4 September 2010.
[277] Rudolf Arnheim, *Visual Thinking* (University of California Press, 1977 [1969]), p. 134.
[278] David Robson, 'A brief history of the brain,' *New Scientist*, p. 24 September 2011. See *Biology Letters*, DOI: 10.1098/rsbl.2011.0570.
[279] Neil Smith, *Chomsky: Ideas and Ideals* (Cambridge University Press, 1999), p. 98.
[280] A. Yuille and S. Ullman, 'Rigidity and Smoothness of Motion: Justifying the Smoothness Assumption in Motion Measurement,' *Image Understanding*, eds. Shimon Ullman and Whitman Richards (New Jersey: Ablex Publishing, 1990), p. 164.
[281] Donald Hoffman, *Visual Intelligence* (New York: Norton, 1998), p. 169.
[282] Maurice Merleau-Ponty, *Phenomenology of Perception*, trans. Colin Smith (London: Routledge, 1962), p. 401.
[283] Graham Lawton, 'The Grand Delusion,' *New Scientist*, 14 May 2011.
[284] Pierre Jacob and Marc Jeannerod, *Ways of Seeing: The Scope and Limits of Visual Cognition* (Oxford University Press, 2003), p. 130.
[285] Ivo Kohler, 'Experiments with goggles,' *Scientific American*, May 1962.
[286] Warren J. Wittreich, 'Visual Perception and Personality,' *Scientific American*, April 1959.
[287] Dennett, *Consciousness Explained*, p. 12.
[288] David Bohm and F. David Peat, *Science, Order, and Creativity* (London: Routledge, 2000), p. 64.
[289] Semir Zeki, 'Art and the Brain,' http://neuroesthetics.org/pdf/Daedalus.pdf, pp. 2-3, originally published in *Daedalus*, 127(2), Spring 1998: 71-103.
[290] James McGilvray, introduction to Noam Chomsky, *Cartesian Linguistics: A Chapter in the History of Rationalist Thought*, 2nd ed. (Christchurch, New Zealand: Cybereditions, 2002), p. 20.
[291] McGilchrist, *The Master and His Emissary*, p. 162.
[292] McGilchrist, *The Master and His Emissary*, p. 162.
[293] John Dewey, *Context and Thought*, University of California Publications in Philosophy, 12(3) (University of California Press, 1931), p. 212.
[294] McGilchrist, *The Master and His Emissary*, p. 330.
[295] Ibid., p. 166 (emphasis his).
[296] Gerald Feinberg, 'Ordinary Matter,' *Scientific American*, May 1967.
[297] Russell, *The Problems of Philosophy*, pp. 46-7.
[298] Bertrand Russell, *Our Knowledge of the External World* (London: Routledge, 1993), p. 151.
[299] Cited in Ian Proops, 'Russell on Substitutivity and the Abandonment of Propositions,' *The Philosophical Review*, 120(2), April 2011: 151-201 (emphasis his).
[300] Davidson, 'The Perils and Pleasures of Interpretation,' *The Oxford Handbook of Philosophy of Language*, p. 1068.
[301] G. K. Chesterton, *Orthodoxy* (1908), ch. 3.

[302] Jeffrey Gray, *Consciousness: Creeping Up on the Hard Problem* (Oxford University Press, 2004), p. 61.
[303] Searle, in Guttenplan, *A Companion to the Philosophy of Mind*, p. 545.
[304] D. W. Winnicott, *Playing and Reality* (London: Routledge, 2005, repr. 2009), p. 93.
[305] Juan Uriagereka, 'Concluding Remarks,' *Of Minds and Language*, p. 406 (379-409).
[306] John Dewey, *Experience and Nature*, 2nd ed. (Chicago: Open Court, 1929), p. 241 (emphasis his).
[307] Huygens, *Cosmotheoros*, p. 57.
[308] D. Premack and G. Woodruff, *Behavioral and Brain Sciences*, vol. 4, p. 515 (1978); Daniel C. Dennett, *Behavioral and Brain Sciences*, vol. 6, p. 343 (1983).
[309] Marc. D. Hauser, Noam Chomsky, W. Tecumseh Fitch, 'The Faculty of Language: What Is It, Who Has It, and How Did It Evolve?', *Science*, 298(5598), 2002: 1575; D. Premack and A. Premack, *Original Intelligence* (New York: McGraw-Hill, 2002); B. Hare, J. Call, B. Agnetta and M. Tomasello, *Animal Behavior*, 59, 2000: 771.
[310] Hauser, Chomsky, Fitch, 'The Faculty of Language,' *Science*, 1576; C. M. Heyes, 'Theory of mind in nonhuman primates,' *Behavioral Brain Sciences*, 21, 1998: 101-14; D. J. Povinelli, T. J. Eddy, 'What young chimpanzees know about seeing,' *Monographs of the Society for Research in Child Development*, 61(247), 1996: 174-189.
[311] Barry C. Smith, 'What I Know When I Know A Language,' *The Oxford Handbook of Philosophy of Language*, ed. Ernest Lepore and Barry C. Smith (Oxford: Clarendon, 2006), p. 953.
[312] See, for example, Ryan McKay, Robyn Langdon and Max Coltheart, "Sleights of mind': Delusions, defences, and self-deception,' *Cognitive Neuropsychiatry*, 10, 2005: 305-26.
[313] McGilchrist, *The Master and His Emissary*, p. 67 (emphasis his).
[314] Michael Ayers, *Locke: Epistemology and Ontology*, 2 vols. (New York: Routledge, 1991), vol. II, p. 285 (emphasis his).
[315] John Muir, *John of the Mountains* (1938).
[316] McGilchrist, *The Master and His Emissary*, p. 438.
[317] Slavoj Žižek, *The Ticklish Subject: The Absent Centre of Political Ontology* (London: Verso, 2008 [1999]), p. 83.
[318] Eric T. Olson, 'Personal Identity,' *The Blackwell Guide to the Philosophy of Mind*, ed. Stephen P. Stich and Ted A. Warfield (London: Blackwell, 2003), pp. 366-7.
[319] Harold Noonan, *Personal Identity*, 2nd ed (London: Routledge, 1989/2003), p. 2.
[320] Timothy D. Wilson's response to V. S. Ramachandran's essay, 'Self Awareness: The Last Frontier,' *Edge*, 1 January 2009, www.edge.org/3rd_culture/rama08/rama08_index.html.
[321] Hume, *A Treatise of Human Nature*, Book I, Part IV, Section VI, 'Of personal identity,' p. 252.
[322] Ibid., Book 1: Of the understanding, Part IV: Of the sceptical and other systems of philosophy, Section VI: Personal identity.
[323] Hinzen, *Mind Design and Minimal Syntax*, p. 9, n. 6.
[324] Ray Cattell, *An Introduction to Mind, Consciousness and Language* (London: Continuum, 2006), p. 1.
[325] Elliott Sober, 'Species Concepts and Their Applications,' *Conceptual Issues in Evolutionary Biology* (MIT Press, 1984), p. 533.
[326] McGilchrist, *The Master and His Emissary*, p. 250. For further discussion see D. Le Bihan, R. Turner, T. A. Zeffiro et al., 'Activation of human primary visual cortex during visual recall: a magnetic resonance imaging study,' *Proceedings of the National Academy of Sciences of the USA*, 90(24), 1993: 11802-5.
[327] Huxley, *Literature and Science*, p. 8.
[328] Aldous Huxley, *Texts and Pretexts* (Chatto & Windus, 1932), p. 5.
[329] Romain Rolland, *Journey Within* (1947).

[330] Marion Milner, *On Not Being Able to Paint* (London: Heinemann, revised edn. 1957 [1950]), cited in Winnicott, *Playing and Reality*, p. 52.
[331] Johann Gottfried Herder, *Sculpture: Some Observations on Shape and Form from Pygmalion's Creative Dream*, trans. and ed. J. Gaiger (University of Chicago Press, 2002 [1778]), pp. 40-1.
[332] Viktor Shklovsky, 'Art as Technique,' in L. T. Lemon and M. J. Reis (trans. and ed.), *Russian Formalist Criticism* (University of Nebraska, 1965), p. 12 (emphasis his).
[333] McGilchrist, *The Master and His Emissary*, p. 393.
[334] McGilchrist, *The Master and His Emissary*, p. 393, quoting Louis A. Sass, *The Paradoxes of Delusion: Wittgenstein, Schreber, and the Schizophrenic Mind* (Cornell University Press, 1994), p. x.
[335] Kenneth Craik, *The Nature of Explanation* (Cambridge University Press, 1943), p. 84.
[336] Drew Leder, *The Absent Body* (University of Chicago Press, 1990), p. 14.
[337] Martin Foss, *Symbol and Metaphor in Human Experience* (University of Nebraska Press, 1949), p. 83.
[338] Antonio Damasio, *The Feeling of What Happens: Body, Emotion and the Making of Consciousness* (London: Vintage, 1999), p. 28.
[339] Ibid., p. 29
[340] McGilchrist, *The Master and His Emissary*, p. 393-4 (emphasis his).
[341] Melvin Konner, *The Evolution of Childhood: Relationships, Emotion, Mind* (Cambridge, MA: The Belknap Press of Harvard University Press, 2010), p. 283.
[342] Nicholas Humphrey, *Soul Dust: The Magic of Consciousness* (Quercus Publishing, 2011).
[343] Guttenplan, *A Companion to the Philosophy of Mind*, p. 51.
[344] David Deutsch, *The Fabric of Reality* (London: Penguin, 1997).
[345] Wartofsky, *Conceptual Foundations of Scientific Thought*, p. 34.
[346] Ibid., p. 36.
[347] Carl Jung, *The Collected Works of CG Jung* (Bollingen Series XX), trans. R. F. C. Hull, ed. H. Read, M. Fordham and G. Adler (Princeton University Press, 1953-79), vol. 10, p. 12, cited in McGilchrist, *The Master and His Emissary*, p. 8.
[348] Jacob and Jeannerod, p. 227.
[349] Ibid., pp. 221-2.
[350] Ibid., p. 222. See J. A. Stevens, P. Fonlupt, M. Shiffrar and J. Decety, 'New aspects of motion perception: selective neural encoding of apparent human movements,' *Neuroreport*, 11, 2000: 109-15.
[351] G. Rizzolatti, L. Fadiga, V. Gallese and L. Fogassi, 'Premotor cortex and the recognition of motor actions,' *Cognitive Brain Research*, 3: 131-41.
[352] McGilchrist, *The Master and His Emissary*, p. 58.
[353] For discussion on how mirror neurons are linked to empathy, see J. Decety and P. L. Jackson, 'The functional architecture of human empathy.' *Behavioral and Cognitive Neuroscience Reviews*, 3, 2004: 71-100.
[354] V. S. Ramachandran, 'Self Awareness: The Last Frontier,' *Edge*, 1 January 2009, www.edge.org/3rd_culture/rama08/rama08_index.html.
[355] V. S. Ramachandran, 'The Neurology of Self-Awareness,' *The Mind: Leading Scientists Explore the Brain, Memory, Personality, and Happiness*, ed. John Brockman (New York: Harper Perennial, 2011), p. 150 (147-52).
[356] McGilchrist, *The Master and His Emissary*, p. 303.
[357] Jacob and Jeannerod, *Ways of Seeing*, p. 46.
[358] Michel de Montaigne, *Essais*, 'Of the power of the imagination,' Book I: 21,cited and translated by McGilchrist, *The Master and His Emissary*, p. 325.
[359] Jean Decety and Thierry Chaminade, 'When the self represents the other: a new cognitive neuroscience view on psychological identification,' *Consciousness and Cognition*, 12(4), 2003: 577-96.

[360] McGilchrist, *The Master and His Emissary*, p. 159.
[361] Max Scheler, *The Nature of Sympathy*, trans. P. Heath (London: Routledge & Kegan Paul, 1954), p. 246.
[362] Llinás, "'Mindness' as a functional state of the brain,' *Mindwaves*, p. 351.
[363] Ibid., p. 358.
[364] Rodolfo Llinás and Denis Paré, 'Brain modulated by Senses,' *The Mind-Brain Continuum: Sensory Processes*, ed. Rodolfo Llinás and patricia Churchland (MIT Press, 1996), p. 6.
[365] Gray, *Consciousness*, p. 3.
[366] Mark Johnson, *The Meaning of the Body: Aesthetics of Human Understanding* (University of Chicago Press, 2007), p. 162. See Vittorio Gallese and George Lakoff, 'The Brain's Concepts: The Role of the Sensory-Motor System in Conceptual Knowledge,' *Cognitive Neuropsychology*, 22, 2005: 455-79.
[367] Johnson, *The Meaning of the Body*, p. 164. See Marc Jeannerod, 'The Representing Brain: Neural Correlates of Motor Intention and Imagery,' *Behavioral and Brain Sciences*, 17, 1994: 187-245; see also S. M. Kosslyn, *Image and Brain: The Resolution of the Imagery Debate* (MIT Press, 1994).
[368] Z. Kourtzi and N. Kanwisher, 'Activation in human MT/MST by static images with implied motion,' *Journal of Cognitive Neuroscience*, 12, 2000: 48-55.
[369] Plato, *Meno*, ed. and trans. R. W. Sharples (Wiltshire: Aris & Phillips, 1985), 85c, p. 79.
[370] Hinzen, *Mind Design and Minimal Syntax*, pp. 68-9, n. 14.
[371] Russell, *An Outline of Philosophy*, p. 118.
[372] George Lakoff, *Women, Fire, and Dangerous Things: What Categories Reveal about the Mind* (University of Chicago Press, 1987), p. 56.
[373] Chomsky, *New Horizons in the Study of Language and Mind*, p. 116.
[374] David Brooks, 'Social Animal,' *The New Yorker*, 17 January 2011.
[375] Noam Chomsky, 'Language and Nature,' *Mind*, 104(413), January 1995: 1-61.
[376] Georges Rey, 'Chomsky, Intentionality, and a CRTT,' *Chomsky and His Critics*, p. 112.
[377] Ibid.
[378] Ludwig Wittgenstein, *Zettel*, ed. G. E. M. Anscombe and G. H. Von Wright, trans. G. E. M. Anscombe (Oxford: Blackwell, 1967), §161, p. 29e.
[379] *The Observer*, 28 August 1955.
[380] McGilchrist, *The Master and His Emissary*, p. 399.
[381] *Annie Hall*, dir. Woody Allen, United Artists, 1977.

5

THE HOLY TRINITY: REFERENCE, REPRESENTATIONALISM, EXTERNALISM

> *A given physical object can be described from the point of view of physics as having a certain length, mass, color, and chemical composition, but a physical description can never tell us if the object is a club or a pole.*
>
> EDWARD SAPIR[1]

With philosophy's 'linguistic turn' well under way, Huxley explained in his classic *The Doors of Perception* that 'our perceptions of the external world are habitually clouded by the verbal notions in terms of which we do our thinking. We are forever attempting to convert things into signs for the more intelligible abstractions of our own invention. But in doing so, we rob these things of a great deal of their native thing-hood.'[2] The obscurity of 'verbal notions' and their connection to Descartes' and Russell's 'objects' has been a major theme of twentieth-century western thought, but as Peter Strawson wrote in a largely ignored and misinterpreted essay on 'referring' and 'presupposition' in 1950:

> Now, whenever a man uses any expression, the presumption is that he thinks he is using it correctly: so when he uses the expression, "the such-and-such", in a uniquely referring way, the presumption is that he thinks both that there is *some* individual of that species, and that the context of use will sufficiently determine which one he has in mind. To use the word "the" in this way is then to imply (in the relevant sense of "imply") that the existential conditions described by Russell are fulfilled. But to use "the" in this way is not to *state* that those conditions are fulfilled. If I begin a sentence with an expression of the form, "the so-and so", and then am prevented from saying more, I have made no statement of any kind; but I may have succeeded in mentioning some one or something.[3]

Strawson identified that 'the use of "the" together with the position of the phrase in the sentence (*i.e.* at the beginning, or following a transitive verb or preposition) acts as a signal *that* a unique reference is being made; and the following noun, or noun and adjective, together with the context of utterance, shows *what* unique reference is being made.'[4] David M. Levin's presents a similar interpretation (reminiscent of schizophrenia), but brings warnings to be heeded later:

> The ego's possibility of mastering, dominating and controlling required, in turn, that objectification – reification – must be given priority, for objectification is the way that the world is brought before us in representation, made available for our technological mastering, and subjected to our domination. But the final ironic twist in the logic of this process of objectification is that it escapes our control, and we ourselves become its victims, simultaneously reduced to the being-available of mere objects and reduced to the being of a purely inner subjectivity that is no longer recognised as enjoying any truth, and reality.[5]

A further example of how language controls our interpretations of reality can be found if we ask the question, 'What makes London, London?' It is comprised of various buildings, roads, houses, universities, parks and so on. These components seem quite necessary and integral to the city. If, for whatever reason, England were to be completely decimated in a nuclear explosion, and the capital rebuilt from the ashes a hundred miles from its previous location, it could still be called 'London,' despite sharing no physical relation with the old capital. London can even be moved without it being destroyed, since virtually everything about 'London' is comprised of abstractions. In fact '*any* polity concept invited both ABSTRACT and CONCRETE characterizations,' as James McGilvray recently put it.[6] Paul Pietroski demonstrates this with the sentence 'France is hexagonal and it is a republic.'[7] Other cases, equally as varied and intricate, may differ:

> If London is reduced to dust, *it* – that is, London – can be *re*-built elsewhere and still be the *same* city, London, unlike my house, which won't be the same house if it is reduced to dust and *it* is *re*-built somewhere else. The motor of my car is still different. If it is reduced to dust, it cannot be rebuilt, though if only partially damaged, it can be. If a physically indistinguishable motor is built from the same dust, it is not the same motor, but a different one. Judgments can be rather delicate, involving factors that have barely been explored.[8]

On the phonetics side, Jackendoff emphasises the subtlety of interpretation in describing neurons with highly selective responses: 'In acoustic phonetics is has been known for a long time that categorical perception of phonemes is not absolutely sharp. Rather, there is a narrow range of acoustic inputs between, say, *p* and *b* that produce variation and context-dependency in hearers' judgments.'[9] McGilvray, developing Chomsky's insights, points out that 'Many nouns have *both* ABSTRACT and CONCRETE features ... if something is a door of wood, it is rigid; but one can also go through the door. A book weighs a kilo and is a compelling story.'[10] A book can also be a stored on a laptop if someone were in the processes of writing one. We will then answer Heraclitus by saying that whether or not one can step in the same river twice depends on which features we are concerned with. But most contemporary philosophers of language would not: Ignoring the crucial point that the conceptual distinction between abstract and concrete features can be found simultaneously in objects, Ernest Lepore writes in an essay on Donald Davidson:

> Tables, chairs, and people are concrete, dated particulars, that is, unrepeatable entities with location in space and time. These features alone distinguish them from, say, either numbers or God.[11]

Pick up your favourite metaphysics textbook and similar confusions abound. Joshua Hoffman and Gary Rosenkrantz, for instance, contend in *The Oxford Handbook of Metaphysics* that 'The distinction between abstract and concrete is, we believe, a fundamental distinction, in that every entity is either concrete or abstract, and no entity is both. We believe that the abstract-concrete distinction is, in fact, the *most fundamental* categorical distinction.'[12] These and other claims ignore the complex abstract-concrete nature of, for instance, 'books.' James Pustejovsky calls such simultaneously concrete and abstract entities 'dot objects,' as in PHYSICAL OBJECT ● INFORMATION.[13] There are many dot-object activities, like reading (which involves the visual system scanning a page and the mind obtaining information from it) and speech acts (which involve the production of noise and the intended transmission of information). This is one of the reasons why nominalist debates are beside the point, focussing (like Aristotle and other pre-Lockean philosophers) on allegedly metaphysical questions, not cognitive ones. The realism versus anti-realism debates in the philosophy of language are also '*metaphysical* at heart,' for Hinzen.[14] Though considering the matter metaphysical and not cognitive or epistemological, D'Arcy Thompson also warned against conflating abstract and concrete features in his classic *On Growth and Form*:

> In Aristotle's parable, the house is there that men may live in it; but it is also there because the builders have laid one stone upon another: and it is as a *mechanism*, or a mechanical construction, that the physicist looks upon the world. Like warp and woof, mechanism and teleology are interwoven together, and we must not cleave to the one and despise the other.[15]

Reviewing recent work in developmental psychology, Lila Gleitman touches on similar concerns:

> Infants generally acquire the word *kiss* (the verb) before *idea* (the noun) and even before *kiss* (the noun). As for the verbs, their developmental timing of appearance is variable too, with words like *think* and *know* typically acquired later than verbs like *go* and *hit*. Something akin to "concreteness," rather than lexical class *per se*, appears to be the underlying predator of early lexical acquisition.[16]

Further research is needed to determine which aspects of dot-objects are innate and which are culturally determined – but, as Jackendoff laments, 'there are just so many goddamned words, and so many parts to them.'[17] Converging our abstract and concrete notions of both 'book' and 'person (author),' Henry James wrote of one of his early books: 'I think of ... the masterpiece in question ... as the work of quite another person than myself ... a rich ... relation, say, who ... suffers me still to claim a shy fourth cousinship.'[18] It's also been claimed by some critics that Wordsworth's publication of *Prelude* helped him overcome his solipsism: Seeing 'himself' concretely before him, embodied by the concrete yet abstract object of a book, similarly blurred the two contradictory notions of 'personal identity.' Other cases defying classical semantic distinctions include when the temperature is simultaneously 90 degrees and rising, and when the average family can have 2.4 children. These and similar constructions 'seem to conflate values and functions, refer to impossible entities, run-together geographical and constitutional conceptions, treat abstract and concrete dimensions on a par etc.'[19]

Taking McGilvray and Pustejovksy's lessons onboard, we could also decide to re-build Babylon, even though the city would bear no material resemblance (or likeness) to the original hanging gardens – contrary to Robert De Niro who, in *The Deer Hunter*, takes a bullet from his gun and states forcefully: 'See this? This is this. This ain't something else. This is this.'[20] Other examples will demonstrate how 'this' is, in fact, many things at once – diverging from the signifier-signified 'Aristotelian tradition' of linguistics and philosophy.[21]

But first, Pierre Jacob develops our understanding of 'reference':

> Let M be the mental representation (or I-construction) associated with the proper name 'London' by the syntax of an I-language. According to the principle of symmetry, M is a member of two distinct relations involving non-mental entities: on the one hand, M can be thought of as an instruction for the articulatory system enabling the pronunciation of 'London.' Thus M stands in relation S, the Sounds relation, with noises of category N. On the other hand, M is in the purported Refers relation R with some presumed nonmental entity E (e.g., a city). Chomsky observes that, unless the nature of entities N and E can be well defined, the relation S and R will remain totally indeterminate. In theoretical linguistics, it is widely taken for granted that a definition of relation S entities N would be of no scientific interest. By parity, and contrary to most externalist philosophers of language, Chomsky concludes that definition of the extrinsic relation R between M and E is scientifically futile.[22]

Undergraduate philosophy discussions (which rarely, if ever, touch upon syntax, treating semantics merely as a sub-chapter of logic), on the other hand, assume that 'a proper name is no more than a "label" attached to a thing: its entire meaning derives from what external thing it is (arbitrarily) associated to, hence it becomes *meaningless* if there is no such external object to which it refers (the so-called "problem of non-existents").'[23] While linguists talk about 'words' (which are used to refer) and 'sentences' (which can be judged true or false), philosophers prefer simply to speak of 'reference' and 'truth' (nouns can refer to objects and sentences can have truth-judgements assigned to them, but not the other way round). Defending his 'convention-based semantics' (which barely scratches the surface of lexical and syntactic complexity), David Lewis wrote that 'Words might be used to mean almost anything ... We could perfectly well use [our] words otherwise – or use different words, as men in foreign countries do.'[24] But this and other variety of 'truth-conditional semantics' holds that sentences are reflections of external 'states of affairs'; sometimes called the 'correspondence theory of truth,' presupposing the existence of mind-independent entities which 'mirror' the semantic terms we use to 'refer' to them, like 'London' or 'cow.' Peter Strawson demonstrates the intuitive sense the correspondence theory makes by asking 'What could fit more perfectly the fact that it is raining than the statement that it is raining? Of course statements and facts fit. They were made for each other.'[25] But nothing in the external world 'corresponds' to the recursive combinatorial construction of linguistic entities through 'narrow syntax.' Even if there were 'mirrored' mind-external entities like London, 'rigidly' existing in the world, this mystical state of affairs would tell us nothing about the nature of our conception of these objects, or how these mental constructions emerged, relate to others, or interact with different cognitive systems.

Psychologist Eleanor Rosch has written: 'Since we understand the world not only in terms of individual things but also in terms of *categories* of things, we tend to attribute a real existence to those categories.'[26] This was well understood by Priestley and the encyclopaedist César Chesneau de Marsais, who warned against interpreting nouns to be 'names of real objects that exist independently of our thought,' let alone 'perceptible objects.'[27] We could also ask what a 'mountain' is, but since the large, protruding mass of land could become an island if the sea levels rose, or a seamount if they rose even higher, we might find it hard to give an answer outside of internalist semantics (the theory that meaning is in the head, not 'out there' in the external world). Strawson also argued that the intention of a speaker determines the meaning of a term and what it refers to (even the 'mythical logically proper name'[28]) – a truism ignored by many philosophers today:

> ...the *expression* [*the king of France*] cannot be said to mention, or refer to, anything, any more than the *sentence* [*The king of France is wise*] can be said to be true or false. The same expression can have different mentioning-uses, as the same sentence can be used to make statements with different truth-values. "Mentioning", or "referring", is not something an expression does; it is something that some one can use an expression to do. Mentioning, or referring to, something is a characteristic of *a use* of an expression, just as "being about" something, and truth-or-falsity, are characteristics of *a use* of a sentence. ...
>
> Meaning (in at least one important sense) is a function of the sentence or expression; mentioning and referring and truth or falsity, are functions of the use of the sentence or expression. To give the meaning of an expression (in the sense in which I am using the word) is to give *general directions* for its use to refer to or mention particular objects or persons; to give the meaning of a sentence is to give *general directions* for its use in making true or false assertions. It is not to talk about any particular occasion of the use of the sentence or expression. The meaning of an expression cannot be identified with the object it is used, on a particular occasion, to refer to. The meaning of a sentence cannot be identified with the assertion it is used, on a particular occasion, to make.[29]

In the light of internalist semantics, the peculiar discussions in much of the mainstream philosophical literature about whether, when Democritus spoke of 'atoms,' he 'literally' referred to 'atoms as we currently understand them' or 'something else,' seem entirely beside the point. Strawson concluded that the 'general moral of all this is that communication is much less a matter of explicit or disguised assertion than logicians used to suppose,' since 'ordinary language has no exact logic.'[30] More generally, Herbert Clark showed in his 1996 study *Using Language* that

> many aspects of live speech that are often taken to be matters of "performance", for example hesitations, repairs, and interjected *um* and *like* are often metalinguistic signals designed to help guide the hearer through the process of interpretation, and to elicit feedback from the hearer as to whether the message is getting across. He [Clark] also shows how gestures, facial expressions, and directions of gaze are often used to amplify the message beyond what the spoken signal conveys.[31]

These debates between 'meaning-as-determined-by-external-factors' and 'meaning-as-determined-by-internal-factors' have come to be known as 'externalism' (meaning is 'outside' the head) versus 'internalism.' Pierre Jacob outlines the difference between the two:

> The debate between internalism and externalism is about whether only brain processes (or processes internal to an individual's brain) underlie a given cognitive achievement or whether properties instantiated in the speaker's environment too are relevant. According to computationalism, the environment may be relevant to fixing the phonological and syntactic parameters of the language spoken around a child. But once fixed, the syntactic and phonological computations operate within the speaker's brain alone.[32]

He also comments on the consequences of internalism for naturalistic inquiry:

> The question is whether from the fact that it is very hard if not impossible to provide necessary and sufficient conditions of individuation for things in the world to be referred to by linguistic expressions, one ought to conclude that there are no things in the world (e.g., cities) for linguistic expressions to refer to and be true of? If so, then it might follow that semantic relations between words and things in the world are not suitable for scientific linguistic purposes.[33]

This has been proven controversial in the philosophical community, with Hilary Putnam's 1975 externalist essay 'The Meaning of "Meaning"' a prime example. Ignoring some of his (misleading) thoughts on the topic of meaning, it's worth noting that he defined water as 'H2O give or take certain impurities.'[34] What defines 'water'? Is there some kind of objective standard from which to determine whether something is 'water'? Chemical composition, it would seem, is the most sensible and irrefutable measure. But on closer inspection 'whether something is water depends on special human interests and concerns.'[35] Though he agrees with these assumptions for different reasons, Mark Johnson also thinks that, 'contrary to the fundamental claim of Gottlieb Frege, the father of modern analytic philosophy, propositions are *not* the basic units of human meaning and thought. Meaning traffics in patterns, images, qualities, feelings, and eventually concepts and propositions.'[36] James McGilvray also observes that 'concepts such as *wish* and *want*, *louse* and *log*, *up* and *underneath*, *good* and *ugly* seem to be virtually designed to serve human interests and needs.'[37] Dewey's later work also stressed the soon to become Wittgensteinian point that and object is 'some element in the complex whole that is defined in abstraction from the whole of which it is a distinction. ... [T]he selective determination and relation of objects in thought is controlled by reference to a situation – to that which is constituted by a pervasive and internally integrating quality.'[38]

If when the 'narrow faculty of language' (FLN) emerged it connected to pre-existing conceptual structures, it follows that non-human primate cognition must enjoy similar 'thoughts' to non-linguistic human 'thought.' Merge, however, seems to be 'narrower' than recursion, since there is 'no clear empirical evidence of recursivity of the structure-building mechanism of grammar in isolation or autonomously from these interfaces [such as the lexicon and planning capacity], notably the syntax-discourse interface'[39] – and so recursion does not exist in the 'narrow' sense of Chomsky, Hauser and Fitch (2002). Berwick and Chomsky comment that conceptual structures are indeed found in non-human primates, including 'probably actor-action-goal schemata, categorization, possible the singular-plural distinction, and others. These were presumably recruited for language, though the conceptual resources of humans that enter into language use appear to be far richer. Specifically, even the 'atoms' of computation, lexical items/concepts, appear to be uniquely human.'[40]

Every single linguistic expression, adds Wolfram Hinzen, carries with it 'massive contextuality.'[41] The water out of an uncontaminated kitchen tap is radically different to the impure substance drunk undernourished children in Uganda. Yet due to human 'interests and concerns' we name both substances water, even if the latter is actually 'H2O plus impurities.' Some European travellers crossing the Congo in 1648 were so famished and starved that they were forced to 'drink water [which] resembled horse's urine.'[42] No doubt a chemist would also reject the notion that the substance from the River Thames is water (or rather 'water defined as H2O'). James Underhill might have had examples like this in mind when he recently wrote, in a study of Humboldt's work, that words 'do not bear any fixed relation to any actual physical thing in the world.'[43] More pertinently, science 'employs whatever concepts and explanatory schemes make sense of the phenomena, and is uncommitted to finding underlying kinds in nouns that we use.'[44] Folk scientific concepts (water) and various assemblies of molecules (H2O) yield separate concepts with different properties. A much basic but crucial objection to externalism is that 'water' and 'H2O' are simply different words, and 'by the general principles of synonymy avoidance that appears to be operative in the human lexicon, they should have different meanings.'[45] In fact H2O 'is not even a term listed in a human lexicon: there is in this sense not even an issue of potential synonymy here.'[46] In the 1930s, Edward Sapir also pointed out that

> a given physical object can be described from the point of view of physics as having a certain length, mass, color, and chemical composition, but a physical description can never tell us if the object is a club or a pole. Two poles may differ greatly in their physical attributes, and a given pole may be identical to a given club from this perspective. What makes us call one a club and the other a pole depends on how we use them or intend to use them, and perhaps our beliefs about the intent of the person who made each object.[47]

In *Vision*, David Marr explored similar problems:

> What ... is an object, and what makes it so special that it should be recoverable as a region in an image? Is a nose an object? Is a head one? Is it still one if it is attached to a body? What about a man on horseback? These questions show that the difficulties in trying to formulate what should be recovered as a region from an image are so great as to amount to philosophical problems. There really is no answer to them – all these things can be an object if you want to think of them that way or they can be part of larger objects.[48]

This certainly applies to simple and complex objects (like 'green' and 'green horse'), but as Marr understood, it doesn't pretend to touch on the internal intricacies and conceptual structures determining the various features (abstract and concrete, for instance) of these objects. What we can be sure of, though, is that 'the notion of an object must satisfy both perceptual and conceptual constraints, since visual perception terminates in recognition and recognition involves the matching of visual information with concepts.'[49] The interest-related and context-bound concerns of Sapir and Marr seem to affect the meaning verbs like 'paint.' If a paint factory explodes and covers parts of the surface of a nearby road with paint, it did not 'paint' the road, so the definition of 'paint' is something like 'intentionally covering a surface with paint.' Akeel Bilgrami and Carol Rovane note the highly contextual and intentional nature of 'reference': 'Social and other external relations do not force a uniform norm of meaning and reference of a term on all speakers of a language, such that all departures from it necessarily amount to mistakes. For some departures, instead of thinking of them as violations of a norm, we think of them as individual ('idiolectical') meanings and references, tied to local contexts of use.'[50] Complimenting Chomsky's stress on 'special human interests and concerns,' they continue: 'There is no question, in any case, of appealing to intentions to refer since we do not and cannot have intentions towards terms in the language of thought, we can only have them towards words we vocalize. Since neither intentions nor conceptions of things play any role, these relations between a term (concept) and an object or property in the world may be the basis of *universal* laws which hold for all speakers who possess the concept and who stand in causal relations with the object or property in question.'[51] Countering the externalist tradition, Chomsky has written that

> the word "water" is a collection of phonetic, semantic, and formal properties, which are accessed by various performance systems for articulation, perception, talking about the world, and so on. If we deny that its meaning is in the head, why not also that its phonetic aspects are in the head? Why does no one propose that the *phonetic content* of "water" is determined by certain motions of molecules or conventions about "proper pronunciation"? The questions are understood to be absurd or irrelevant. Why not also in the case of meaning?[52]

In work that will surely become a standard text in biolinguistics, Hinzen makes the similar insightful point that 'we haven't got sense organs for [meanings]' in the same way we do for sounds.[53] Denouncing the 'prevailing externalist orthodoxy,' he adds that 'just as the study of our "bodily" organization has a fundamentally *internalist* aspect – what organs mature in ontogeny is a function of genetic factors and laws of development – it assumes that the study of our "mental organization" should have this aspect too.'[54] The very notion of externalist semantics, then, has about as much coherence as externalist syntax. Contemporary philosophy of mind is also riddled with similar confusions. Take Andrew Melnyk's revered *Physicalist Manifesto*, in which he claims that

> Someone with the concept of table salt could gain competence with the concept of NaCl in a chemistry class that failed to mention that NaCl is the same stuff as table salt. It would then be conceivable for this person, in the sense specified, to believe that his food could be sprinkled with salt but not with NaCl: he could entertain the thought that this is so without explicit contradiction or any sense of conceptual blockage. But it is not really possible that his food could be sprinkled with salt but not with NaCl, since salt and NaCl are the very same stuff and it is not possible that his food be both sprinkled and not sprinkled with the very same stuff. So the conceivability, in the sense specified, of a proposition does not in general entail its possibility.[55]

Paul Pietroski, in a recent collection of essays on contextualism, will certainly have none of this:

> The environment is surely responsible for some aspects of intersubjective stability with regard to how we *use* linguistics expressions to talk about things; although this kind of stability may well presuppose meaning. Perhaps serious investigation will eventually reveal that our shared understanding of the word "water" has more to do with H2O than with internalistic properties of the world.[56]

And if our understanding of H2O does in fact rely heavily on molecular structure, then so be it. The facts will be whatever they turn out to be. Based on the evidence presented so far, however, it's hard to deny the overwhelming importance of cognitive and environmentally contingent factors on the complex and intricate activity of human concepts. Pietroksi's sentence 'France is hexagonal and is a republic' again demonstrates the effortless teamwork between concrete and abstract lexical features. Departing radically from Kripke, he also elaborates on how expressions constrain but do not determine truth-'conditions':

> As a matter of causal-historical fact, speakers use this "famous-person name" [Aristotle] to talk about a certain long dead philosopher. Speakers also tacitly know that names have both causal-historical associations and descriptive associations; that

these aspects of use can conflict, with regard to 'who we are talking about' when use that name; and that in such cases, the causal-historical associations trump the descriptive associations. But the plausible hypothesis that names are devices for referring "rigidly" (and not by description) does not require the implausible hypothesis that names denote things. On the contrary, one might ask which thing "France" rigidly denotes. For we can coherently describe a possible situation in which the terrain of France is not inhabited by people who have a republican form of government, and a possible situation in which the republic of France has different borders.[57]

Again to echo Wittgenstein (who was himself influenced by Frege's 'context principle' in his *Grundlagen*; that only within a construction does a word have its meaning), the terms 'water' and 'H2O' are learnt in strictly different ways, with 'water' being a common sense term and 'H2O' being a technical term (which cognitive capacities enter into the different realisations of such concepts is yet to be determined). We learn 'water' in childhood English lessons, but not 'H2O,' which we pick up later in chemistry textbooks. Neither do chemists learn the meaning of the word 'water' in university (though they may use it informally). So the sentence 'water is H2O' is simply a meaningless confusion of two separate languages, making the answer to Juliet's question, 'What's in a name?' less simple than it appears. A 'circle,' for instance, is a common-sense, folk-scientific notion, but an 'ellipse whose foci coincide' is not. We can consequently equate 'meaning' with 'perspective': the 'meaning' of 'water,' contrary to Putnam, depends on whether we take a 'cognitive' or 'physicalist' perspective. But this is not quite right, since the actual *concept* of water is not reducible to motions of molecules, contrary to dominant illusions. This is one of the reasons why Nelson Goodman used to advise speaking not of the 'concept of X' but instead 'X-concept,' to avoid what today would be called externalist intuitions. Unlike *water*, '*English* and *sunset*, words like *puddle*, *laundry*, *junk*, and *buddy* don't lend themselves to perspectives other than the ordinary one.'[58] Theoretical terms like 'X-bar' also lack an ordinary conceptualization. Furthermore, learning terms like 'H2O' compels us to reflexively nestle the concept to more familiar idealisations, like 'water': "*Knowing*' is how we come to feel that we *already know* something: thus, it means *combating a feeling of newness and transforming the apparently new into something old.*'[59] What Priestley believed to be the 'essence' of an object, having been 'superadded' by God, is in fact added by the mind. Our folk-scientific notion of 'water' (no doubt similar to Aristotle's) plays no part in a chemist's understanding of H2O, since 'virtually nothing we say about the 'physical world' ... can be translated into the sciences.'[60] In fact, quite surprisingly, 'Experiments by Barbara Malt show that *water* – even *prototypical water* – correlates quite weakly with H2O content, even for people who know the relevant chemistry.'[61] A report in *Science* also points out that 'Most of the universe's water exists in the glassy state.'[62]

We are often met with similar confusions in talking about traditional philosophical questions from a naturalistic point of view. Dawkins, bearing the torch of Reason as always, believes that life has no 'meaning' and 'purpose.'[63] But the common-sense notion of 'the meaning of life' is easy enough to grasp (as is the concept of the 'English' language), and saying that it doesn't exist is, again, simply conflating the perspective of common-sense notions with the perspective of scientific claims about the nature of organic life, matter, and so on. The world doesn't suddenly change when you walk into the physics department, rather the perspective changes. Dawkins doesn't even attempt to give an explanation for the existence of concepts like 'the meaning of life' – a much harder task. As he often reminds his teenage Christian rivals during lecture time, dangerous and irrational belief systems have no place in a scientific community.

Francis Crick famously claimed that 'you are nothing but a pack of neurons ... You, your joys and your sorrows, your memories and your ambitions, your sense of personal identity and free will, are in fact no more than the behaviour of a vast assembly of nerve cells and their associated molecules.'[64] But this view, like many others, drastically conflates the 'abstract' notion of person and the 'concrete' notion of body ('I' am not my 'body' and my 'body' is not 'me'). Internalist semantics also renders Searle's distinction between 'micro-elements' and 'macro-properties' incoherent[65], along with phrases like 'the molecules of water.'[66] The 'water' in a highly polluted river and the force-dynamics of 'motion' do not suddenly disappear when theories of molecules and gravity appear, and so 'commonsense can continue on its merry way.'[67] Folk science is impossible to be dismissed by semanticists, as Pinker argues:

> Our words and construction disclose conceptions of physical reality and human social life that are similar in all cultures but different from the products of our science and scholarship. ... Our ability to combine them into bigger assemblies and to extend them to new domains by metaphorical leaps goes a long way toward explaining what makes us smart. But they can also clash with the nature of things, and when they do, the result can be paradox.[68]

Acknowledging that there are various conflicting perspectives we can take to interpret the world (physicalist, folk, etc.) doesn't mean, as is sometimes claimed, that everything is relative, but rather that there are many absolutes. This talk of 'perspectives' is a form of projectivism (as the term is understood in the visual and cognitive sciences). Interacting with the C/I interface, the language faculty uses 'the conceptual resources that are available. ... It's a little hard to say what language is "itself." Does the English language include the word *gravitation*? We're somehow in an area now where our linguistic capacities and our science-creating capacities are interacting. We don't understand either of these systems enough to know where to go on.'[69] McGilvray suggests we think of lexical items as 'the language faculty's contribution to a human's conceptual (configurational) capacities.'[70] In *Daybreak*, Nietzsche displayed his philological side in pointing out that 'In earlier times, it was ... concluded that where the realm of words ceased, the realm of existence ceased also.'[71] But '[t]he contribution [of language to human concepts] is partial,' so 'language' is indeed 'thought,' just not all of it.[72] As Drummond and Hornstein speculate on the relationship between FLN and the C/I interface, 'the bit that does the merging and moving will turn out to be a lot smaller than the bit that does the thinking.'[73]

The above internalist assumptions seems to corroborate Sapir's belief that 'No entity in human experience can be adequately defined as the mechanical sum or product of its physical properties. ... [I]t is notorious how many of these physical properties are, or may be, overlooked as irrelevant.'[74] Writing on artistic imagination, Antoine de Saint Exupéry hinted at something similar: 'A rock pile ceases to be a rock pile the moment a single man contemplates it, bearing within him the image of a cathedral.'[75] Unlike water, H2O is not an 'object' in the cognitive sense (not much comes to mind when we see a drawing of two hydrogen atoms and an oxygen one), it's instead a posited physical entity. In his internalist study *The Stuff of Thought*, Pinker observes that 'modern scientific definitions for a word may be at odds with the way that people (including the scientists themselves) used to think about its referent. Whales used to be thought of as very large fish (the animal that swallowed Jonah was called "a big fish" in the original Hebrew), but today we know they are mammals. But when scientifically illiterate people use the word *whale*, they are surely referring to the same animals that we are, not to some fish that happens to be large, like a whale shark.'[76] Whatever conceptual tools, sounds and images are employed by the mind in its attempt to grasp 'H2O' or 'mammal' is irrelevant to the physicist, though it may of some interest to the ethnoscientist. Consequently, the word object has an unavoidably 'chair-like feel,' notes Galen Strawson: 'Our most fundamental thought categories simply do not get the world right, and when we think hard we can see a priori that this is so.'[77]

As Philip Carr puts it, folk science and natural science 'exist side by side, offering two completely different pictures of the world.'[78] Along with Sapir, Eddington noticed that there are two kinds of tables: the ones present in ordinary experience, and the ones composed of atoms in space-time (though even this stance equates the concept 'table' with some mind-external entity, instead of simply the world, which remains in Lange's 'mystic obscurity').[79] Hinzen follows Chomsky in pointing out how 'physics has discarded ordinary notions of movement, causation, or solidity, when talking about the physical world. Tables and chairs, too, do not figure in physical theories, and no physical theory can account for what it is for something to be a table, in our understanding.'[80] As with Sapir's club and pole, 'no amount of physical description of some object (in terms of quarks, strings, electromagnetism, and so on) would ever reveal what is most obvious to *us*, as we look at it: say, that is a *table*.'[81] In his account of 'theoretical objects' (those entities posited to be 'real' by scientific theories), Gordon Brittan Jr. notes the following concern:

> Science is made possible by the introduction of theoretical objects. Why this should be so has never been made clear. Indeed, it has never been made clear how theoretical objects are rightly to be understood, or in what ways they differ from more ordinary sorts of physical objects.[82]

The most useful distinction to adopt between these perspectives, I think, is Ludlow's categories of I-substance ('intuitive,' like water) and P-substance ('physical,' or theoretical, like H2O).[83] A separate question emerges from the ethnoscientists over what cognitive capacities enter into the structuring of both concepts (if H2O even qualifies as a concept), questions which may in fact reveal the limits imposed by our conceptual structures on naturalistic inquiry.

A view emerging from Wittgenstein's *Philosophical Investigations* is that linguistic expressions have no fixed meaning – a cautiously generalised partial truth. But, to no surprise, context and human intention are certainly not the only factors to consider, since an 'expression must convey *something* with which the context can interact. If it did not, a hearer could in principle know from the context what message was intended, without the speaker saying anything at all!'[84] The inherent 'indirectness' (or 'anti-realist' as it's sometimes misleadingly called) nature of this internalist stance was often hinted at by Russell, for whom, famously, descriptions are 'incomplete symbols'; phrases which 'have no meaning in isolation' but are only 'defined in certain contexts.'[85] Discussing Russell's essay 'On Denoting,' Thomas Baldwin draws the following conclusions:

> For once one adds, as Russell does here, that we are not acquainted with matter and the mind of other people, it follows that our understanding of sentences that include putative names of material objects and other people cannot be achieved by identifying these things as the meanings of the names; instead we are bound to reinterpret the names as descriptions that invoke properties of things with which we are acquainted.[86]

Though this doesn't mean, as Baldwin implies, that 'putative names of material objects and other people' are merely descriptions, since this would still presuppose some level of direct acquaintance. Putative statements are descriptions of people and objects to the same extent that metre readings of photon activity at the Large Hadron Collider are descriptions of 'matter.' Even in the case where the chemist would not be able to distinguish between two substances, we can still choose to differ in terminology. If, for example, I insert a tea-bag into a mug of water, and put beside it a measure of liquid taken from the Thames which happens to have been contaminated by tea leaves floating in the river, are these indistinguishable substances both 'water'? The answer is down to individual choice (similar games can be played with 'chalk' and 'CaCo3'). If someone at an ice bar saw a glass-shaped piece of frozen ice with some water and ice cubes in it, and asked the barman for some water, he'd be surprised it the barman poured the water out and handed him the ice cubes. Confusions would mount if he asked for a 'real' glass of liquid, before watching the barman hand him an empty glass, with the chemist sitting next to him pointing out that glass has molecular characteristics which technically qualify it as a liquid (they are actually amorphous solids, but share many structural properties with liquids). The results of these semantic exercises seem reminiscent of Locke's claim that 'Words, in their primary or immediate signification, stand for nothing but *the ideas in the mind of him that uses them.*'[87] The fourteenth-century French philosopher John Buridan also held a similar internalist view in his *Summulae de dialectica*: 'the capability of speaking was given to us in order that we could signify our concepts to others and also the capacity of hearing was given to us in order that the concepts of speakers could be signified to us.'[88]

Chomsky cites earlier sources to strengthen the insight that thoughts are not 'external objects':

> The basic assumption that there is a common store of thoughts surely can be denied; in fact, it had been plausibly denied a century earlier by critics of the theory of ideas who argued that it is a mistake to interpret the expression "John has a thought" (desire, intention, etc.) on the analogy of "John has a diamond." In the former case, the encyclopedist du Marsais and later Thomas Reid argued, the expression means only "John thinks" (desires, etc.) and provides no ground for positing "thoughts" to which John stands in relation.[89]

Though his suspicious gaze was cast primarily on eloquence, Francis Bacon's remarks could easily be seen as a valuable lesson to contemporary philosophy of language and mind: '[M]en began to hunt more after words than matter; and more after ... tropes and figures, than after the weight of matter ... [and] soundness of argument.'[90] I can't help noticing the irony in the fact that I came across this quote in an introduction to literary 'theory,' whose authors seemed oblivious to its import. Even more oblivious to the insights of linguistics was an English graduate from the University of York I once met who, after I asked him whether he thought 'meaning' was in 'the text,' 'the head,' 'the author' or 'the world,' told me 'I think the text has some work to do' (for similar literary and cultural 'theorists' 'the text' is to be regarded as an object of profound, even mystical power, to be regarded with fear and awe, and not simply a product of behaviour, like cave etchings or an artist's canvas). The above uncontroversial assumptions, then, make the 'reader response theory' in literary criticism (the view that, contrary to Leavis, there is no single 'objective' meaning to a text, and that every reader interprets 'the text' differently) seem slightly less radical and inflated than Barthes and many English lecturers claim (it speaks with an air of 'words' over 'matter,' to repeat Bacon). Likewise, many terms in philosophy, like 'representation' and 'content,' hang on like a thread by having their non-technical definitions evoked in the mind of the reader instead of their intended technical sense. Elsewhere, Bacon wrote that 'Words still manifestly force the understanding, throw everything into confusion, and lead mankind into vain and innumerable controversies and fallacies.'[91] If the incoherent technical notion behind the term 'reference' were to be given a new title of an obscure Greek origin, no doubt it would find it harder to survive.

But nothing remotely like a referentialist relation exists for human language. Animal communication, on the other hand, seems to be grounded in some sort of relation between a sound and an extensionally and physically identifiable entity (as Nirmalangshu Mukherji recently argued).[92] The leading observer of wild chimpanzees, Jane Goodall, notes that for them 'the production of a sound in the *absence* of the appropriate emotional state seems to be an almost impossible task.'[93] It is therefore quite plausible, as Gallistel has shown, that non-human animal cognition can be best described as a non-intentional referential relation.[94] Specific monkey calls also have relations to external events ('The rain is falling') or internal states ('I'm hungry').[95] Non-human animals seem unable to free themselves from these externally-determined constraints, since 'all of the putatively referential calls they produce map onto a narrow range of actually present objects or events, even though these animals confront a wide range of social and ecological situations that are, from a functional perspective, *worthy* of comment.'[96] Quine famously called language 'a social art which we all acquire on the evidence solely of other people's overt behavior under publicly recognizable circumstances. Meanings, therefore, those very models of mental entities, end up as grist for the behaviorist's mill.'[97]

This externalist analysis of meaning, in which 'mind-external and language-independent 'propositions' are abstract Platonic objects that have their truth conditions essentially,' assumes that natural language has 'those truth conditions only 'derivatively.''[98] This is 'hard to square with the internalist viewpoint that the biolinguistic program has adopted.'[99] Reference is consequently a fine topic for neo-Gricean pragmatics, but not semantics. It's no surprise that the origins of biolinguistics can be traced to Descartes (or even to the Medieval grammarians), whose ideas were revived in linguistic departments by Chomsky's *Cartesian Linguistics*, and who argued that in contrast to animal 'language,' human words 'make no reference to any passion' with the exception of emotional outcries (a prototypically internalist view).[100] Considering the differing 'mental' architecture of human and other great apes, Maggie Tallerman has recently followed an implicitly Cartesian line in suggesting that, 'Unlike mere *categories*, which other animals do have, concepts involve offline thinking (thinking about an activity or entity you are not currently engaged with) and displacement (e.g. imagining a leopard that's not present). Concepts have permanent storage in the brain which can be accessed voluntarily (a lexicon).'[101] These lexical items and concepts are not to be found on the beaches or in the forests, leading Galen Strawson to call them 'I-concepts' (though it's unfortunate that he needs to stress the organism-internal nature of concepts, a reminder of the grip of the referentialist religion). Discussing the externalist/internalist distinction, McGilvray notices that

> One can to a degree "determine" a referent with respect to a context by including in an account of "what is said" a set of indices that specify time of speech, person speaking, context of speech, purpose of speech, and so on. Or rather, one can hope that something like this would succeed, for it surely cannot: the

indices would have to be endless in number to cover all possible cases, and there would be no discourse-independent means of fixing even a single set of assignments, assuming – counterfactually – that they would suffice.[102]

Internalist studies of the mind are 'not concerned with mental *representation* – as a relation between mind and world, as a function that the mind carries out – but with mental *representations*, considered as specific natural objects that enter into language use and are invoked to explain it.'[103] As the judicious scholar John Yolton puts the matter in an essay on Rorty: 'The representative function of ideas is not a sign function: ideas are not signs of things, they are the interpretations of, or cognitive response to, physical motion' – where the final two words for biolinguistics simply refer to 'the world,' with the terms 'physical' and 'motion' maintaining only their common-sense definitions and not their Cartesian metaphysical ones.[104] A further absurdity in the mind sciences and philosophy, as Gilbert Harman sees it, is that 'while hardly anyone thinks one can do philosophy of physics without knowing physics, it is all too common to think one can do philosophy of language without knowing linguistics.'[105] Identifying widespread double standards (or 'methodological dualisms') in the study of the mind, Chomsky and Berwick make the following observations:

> Being reflective creatures, thanks to the emergence of the human capacity, humans try to make some sense of experience. These efforts are called myth, or religion, or magic, or philosophy, or, in modern English usage, science. For science, the concept of reference in the technical sense is a normative ideal: we hope that the invented concepts *photon* or *verb phrase* pick out some real thing in the world. And of course the concept of reference is just fine for the context for which it was invented in modern logic: formal systems, in which the relation of reference is stipulated, holding for example between numerals and numbers. But human language and thought do not seem to work that way, and endless confusion has resulted from failure to recognize that fact.[106]

Although 'the theory of reference has formed almost an axiomatic basis for much of the philosophy of language and the philosophy of mind, it has not been systematically investigated empirically or experimentally tested,' Hinzen adds. This was not relevant to Frege and Russell's logic paradigm, as Chomsky notes, but within biolinguistics it is 'the very point.'[107] Unlike the sentences in Tarski's formal languages, natural language 'sentences include expressions whose contributions to what is said using them are determined only relative to contextual factors.'[108]

Today, the claims of contemporary philosophers of language should recall Chomsky's brief comments in 1957 on the relationship between scientists and philosophers; that 'A scientist concerned with meaning (or use) faces the problem of justifying somehow the descriptive statements that he makes. The insights of philosophers into ordinary language will appear to him as posing problems that he must ultimately solve, on some objective basis.'[109] When dealing the work of externalist philosophers, the lack of empirical support or any 'objective basis' for their claims should persuade us to take Hume's advice:

> When we run over libraries, persuaded of these principles, what havoc must we make? If we take in our hand any volume; of divinity or school metaphysics, for instance; let us ask, Does it contain any abstract reasoning concerning quantity or number? No. Does it contain any experimental reasoning concerning matter of fact and existence? No. Commit it then to the flames: for it can contain nothing but sophistry and illusion.[110]

It is sometimes hard to resist our externalist intuitions, as when Galen Strawson confuses 'the world' with 'external object': 'For an external object is by hypothesis an essentially non-mental thing, and is *obviously* of an entirely different species from an essentially mental thing like an impression and an idea.'[111] Carl Juhl and Eric Loomis have recently introduced the term 'analytic*' to stand for a sentence-as-used and understood by a linguistic community, 'a common form of stipulation,' though tantamount to a new construction of the E-language concept of 'public languages.'[112] Falling into this E-language trap, Charles Barber also wrote in his standard history of English that 'a language is something which is *spoken*.'[113] Biolinguists interpret concepts differently from Davidson, for whom 'Concepts themselves are abstractions and so timeless.'[114] They also differ from Fodor on his fifth 'non-negotiable' requirement for concepthood, that 'Concepts are *public*; they're the sorts of things that lots of people can, and do, *share*.'[115] The distinguished philosopher Bill Brewer has also recently sided with 'the realist contention that these direct objects of perception are the persisting mind-independent physical objects we all know and love.'[116]

Almost anticipating Putnam's externalist 'direct realism'[117], McDowell also proposes the mind's 'direct openness to the world.'[118] David Lewis has also written that 'meaning' is to found in the E-language notion of 'conventions' or regulated forms of verbal behaviour.[119] But, drawing on Pietroski, these 'conventions' do not explain why 'The senator called the millionaire from Texas' can be interpreted as a call from Texas, or the millionaire from Texas, but not the senator from Texas.[120] Only an internalist (and, indeed, biolinguistic or 'Galilean') exploration of syntactic processes can explain this phenomenon. We can then read Lewis and safely 'know that he is a member of the Humpty Dumpty army and be very careful with him.'[121]

Even the recent development of 'two-dimensional' semantics (which distinguishes between a word's primary and epistemic intension) presupposes quite enthusiastically a word-object reference relation, in which water is treated as 'a singular term referring to a chemical kind.'[122] Two-dimensional semantics works on the assumption that, 'Since it is not a priori that water is H2O, there is an epistemic possibility in which water is, say, XYZ, and not H2O, even though there is no such metaphysical possibility.'[123] But this is clearly cannot be the case (as Jackendoff might say, 'There is absolutely nothing right here'), since the concept of water carries no ontological presuppositions, unlike H2O. François Recanti also speaks to the representationalists when he claims that 'we can think of concepts ... as mental files in which we store information concerning the extension of the concept.'[124] But there exist no extensions of concepts which these 'mental files' pick out of the world; there are simply concepts or representation, neither of which are *of* anything except themselves (making talk of 'reference via causal-historical chains' incoherent).[125] In a recent essay on 'the origins of meaning,' the distinguished linguist James Hurford slips into similar puzzling territory in claiming that, 'In human language, words stand for concepts, and concepts themselves (at least the concepts of concrete things) relate to entities in the external world in roughly the same way that non-human animal proto-concepts do.'[126] It is one of the tasks of biolinguistics to resist these intuitions, explaining instead how meaning is couched *as though it were external*. Externalists and many other philosophers, fanatically waving their versions of the Twin-Earth or Dry-Earth scenarios (along with Dretske's zoo puzzle), avoid the much harder task of exploring the architecture of our conceptual structures, even putting aside the glaring (and non-trivial) question of why it is we seem to intuitively equate water with H2O in the first place. I think, then, that most of the work done in the philosophy of language over the last half century by the likes of Kripke, Putnam, Chalmers, Davies, Humberstone, Stalnaker, Jackson, Kaplan, LaPorte and Soames is either an indirect contribution to syntactic theory or essentially worthless.

As with the differences in reference relations, non-human animals also see a picture of a plant as an actual plant, not as a representation of one, evidence suggests. As for the human visuomotor and visual perceptual systems, Jacob and Jeannerod argue for what they call the 'representational theory of the visual mind,' which stipulates that 'mental processes consist of the formation and the transformation of mental representations.'[127] These, they stress, are internal representations, not representations *of* extra-mental objects. Non-mental representations include cultural artefacts and other products of behaviour, such as texts, 'linguistic utterances, mathematical and logical symbols, diagrams, road-signs, maps, states of measuring devices (e.g. gauges, thermometers, scales, altimeters, etc.), paintings, drawings, photographs and movies.'[128] Though the human brain lies in far greater obscurity, the 'basic contrast' between the temporal and parietal visual associative areas in a group of monkeys studied by Mortimer Mishkin 'is drawn in terms on two kinds of visual attributes of a perceived object: its intrinsic qualities and its extrinsic spatial relationships to other objects in the visual array.'[129]

The notion of 'object' can ultimately only be understood from an internalist perspective (adding the word 'external' doesn't allow one to get away with externalist presuppositions). Hume goes slightly further, but not far enough: 'The farthest we can go towards a conception of external objects, when [they are] suppos'd specifically different from our perceptions, is to form a relative *idea* of them, without pretending to *comprehend* the related objects.'[130] He does, however, later make this crucial point: 'Since we may suppose, but never can conceive a specific difference betwixt an object and impression; any conclusion we form concerning the connexion and repugnance of impressions, will not be known certainly to be applicable to objects.'[131] There is, he adds in the *Enquiry*, 'no known connexion between the sensible qualities and *the secret powers* [of bodies].'[132] Correspondingly, 'no philosopher, who is rational and modest, has ever pretended to assign *the ultimate cause* of any natural operation, or to show distinctly the action of *that power, which produces* any single effect in the universe. ... [B]ut as to *the causes of these general causes*, we should in vain attempt their discovery. ... These *ultimate springs and principles* are totally shut up from human curiosity and enquiry.'[133]

Kripke's proposal in his early essays that a table is *necessarily* a table, picking out the same extension in every possible world even if it's occasionally used for various other purposes than to dine on, presupposes that the identity of the table is externally determined by, presumably, natural laws similar in scope and origin to those which determine at what temperature H2O boils at certain altitudes. In his *Consequences of Pragmatism*, Rorty gave the following account of Kripke's externalist physicalism:

> Whereas Frege, like Kant, thought of our concepts as carving up an undifferentiated manifold in accordance with our interests ... Kripke sees the world as already divided not only into

particulars, but into natural kinds of particulars and even into essential and accidental features of those particulars and kinds. The question "Is 'X is Φ' true?" is thus to be answered by discovering what – as a matter of physical fact, not of anybody's intentions – "X" refers to, and then discovering whether that particular kind is Φ. Only by such a "physicalistic" theory of reference, technical realists say, can the notion of "truth as correspondence to reality" be preserved.[134]

There is not much sense to be made from Kripke's externalist semantics here (or Carl Hempel's, whose 'empiricist criterion of meaning' honoured the spirit of logical positivism by claiming a statement to be meaningful if and only if it admitted itself to experiential test). The term semantics, it's worth noting, should really be regarded as a form of syntax, since it's the study of internal representations: As Chomsky points out, 'It is possible that natural language has only syntax and pragmatics; it has a "semantics" only in the sense of "the study of how this instrument, whose formal structure and potentialities of expression are the subject of syntactic investigation, is actually put to use in a speech community."'[135] It's of no surprise, then, that the word 'glamour' derives from 'grammar.'[136]

Even mountains could become islands if the sea levels rose high enough, making them not *necessarily* anything. But these externalist assumptions originate further back than Kripke, as 'We have Sokrates to thank for the phantom of definitions which presuppose an altogether imaginary agreement of name and thing.'[137] Lange's other caveat has also not been heeded by mainstream philosophy of language:

> So soon as the words are treated as adequate pictures of things, or are confounded with real picturable things, while they are really only arbitrary signs for certain ideas which must be used with great care, the field is opened to innumerable errors.[138]

The semantics of Kripke and Putnam also completely contradict the way we use language. Putnam's externalism effectively claims that after a linguistic expression is uttered the world takes over (so to speak) and determines its referent. It's not too difficult to demonstrate how this Millian proposal that the meaning of a term is its referent is clearly incoherent. Russell's claim that the identity of a nameable 'object' or 'thing' is contingent on spatio-temporal contiguity was also false. To use one of Chomsky's examples, a collection of leaves on a tree is not a nameable object, but if human choice and action were behind their appearance on or around the tree for the purposes of decoration we could identity them as an object. The four legs of a cow could also be seen as a single object under many conceivable contexts, such as if they were cut off, tied together and used as a doorstop, still being understood by its user to be part of a cow. Abstract objects, then, do not have causal relationships and are not spatiotemporally located (an 'object' is also usually understood to be a concrete thing, hence the confusion when some are denied spatiotemporal relations). Instead of letting the world take over, we should recognise with Hinzen that 'meaning is a structural and internalist phenomenon, relating to the emergence of order and of complex organization in the human language faculty, and other cognitive systems inside the mind interfacing with it.'[139] Meaning never leaves the head: It is not 'preserved, or transmitted, between speakers and over time,' contrary to mainstream dogma.[140] '[R]eference-fixation is an *a priori* matter,' contra Putnam, whose fellow externalists speak of objects existing 'out there' in the external world.[141] But as with physicalist terminology, it is far from clear what 'out there' means:

> Outside our bodies is a physical world, but no abstract properties: scientists study phenomena, usually unspecified for ontological category, and hypothesize them to have certain abstract properties that explain their behaviour. Positing these properties is based on explanatory constructs that scientists creatively come up with in their heads. For these explanations to work we need no ontology of external objects called "properties" that one-to-one correspond to these explanatory constructs. No scientist studies "properties", in the sense of natural objects; they could not, I suppose, if these are abstract objects.[142]

Philosophers Karola Stotz and Paul Griffiths agree that we should think of scientific concepts 'as tools of research, as ways of classifying the experience shaped by experimentalists to meet their specific needs. Necessarily, these tools get reshaped as the demands of scientific work change.'[143]

As if often the case, the answer to the question of the identity of objects 'out there' is really to be found 'in here' (as one of Peter O'Toole's characters confessed: 'I find that when I pray, I'm talking to myself'). 'The problem of the external world' (as Ram Neta and others put it) then becomes not a purely metaphysical question, but a cognitive and naturalistic one.[144] Scientific 'facts' are consequently 'a *consensus* of rational opinion over the widest possible field,' and not a matter of empirical certainty, leading Socrates to often complain about the lack of unity between the numerous Athenian schools of natural philosophy.[145] This is one of the number of reasons Lewis Mumford thought 'astrophysicists ... must reckon with ... the possibility that their outer world is only our inner world turned inside out.'[146] For Mark Johnson, 'Objects, as we experience them, are actually stable affordances for us – stable patterns that our environment presents to creatures like us with our specific capacities for perception and bodily action.'[147] Favouring the above Hobbesian and Chosmkyan thought experiments, Hinzen believes 'An object shows its true nature not by exploring its natural and holistic connections with other things, but by putting it in an artificial circumstance where it establishes accidental and "unnatural" connections with other objects.'[148] Russell also saw the crucial importance of such an approach to unearthing the hidden structure of logic and semantics, claiming in his early work that 'A logical theory may be tested by its capacity for dealing with puzzles, and it is a wholesome plan, in thinking about logic, to stock the mind with as many puzzles as possible, since these serve much the same purpose as is served by the experiments in physical science.'[149] The biolinguist will therefore answer Carl Hempel's puzzle of how the statement 'All ravens are black' could be refuted by discovering a raven of an alternate colour (would it still be a raven, he asks?), by pointing out that 'raven' is an arbitrary taxonomic classification, not a metaphysical term.[150] Darwin also held that a term for a species should be 'a pointer to a population of organisms (a rigid designator) rather than as a type that is stipulated by a fixed set of traits.'[151] Though this may frighten some biologists, Francis Crick noted the difficulty for 'many people to accept that what they see is a symbolic interpretation of the world – it all seems so like "the real thing". But in fact we have no direct knowlege of objects in the world.'[152] Semantic vagaries also abound when trying to determine 'objecthood': Tom Bever has shown, for instance, that some adjectives are more noun-like than others; 'red' can be a colour, but 'nice' can't be anything.[153] In fact the only sure way to find out what an object is (and to answer Davidson's question 'What is it for words to mean what they do?') is to explore the mental faculties which constantly trick us into defining and re-defining reality.[154]

Jackendoff's infamous speculations on a distinct semantic or 'conceptual' structure to account for certain peculiarities in language (such as why 'The trolley rumbled around the corner' is understood to mean that the trolley moves, despite the fact that the verb is one of sound emission, not motion) results from him charging biolinguistics of being 'syntactocentric.'[155] But his criticisms are based on a misinterpretation of minimalism, which claims that FLN, recursion, is 'autonomous' from other cognitive systems. As Hinzen explains, 'Only on a very superficial understanding does the thesis of the "autonomy of syntax" mean that generative grammar is "not concerned with meaning". Chomsky for one never held this view. In generative grammar, it is expressions-under-an-interpretation (forms with meaning that depend on these forms) that are to be derived.'[156] If the charge of syntactocentrism were at all meaningful, 'it would seem that phonologists could level a similar charge, but they rightfully do not. ... [U]nderstanding how humans use language meaningfully and understand expressions clearly *depends* and is *explained* by internal mechanisms that make these overt effects impossible.' The 'optimal design' case (or the Strong Minimalist Thesis (SMT) that FLN is optimally designed to meet the demands imposed by the cognitive systems it interacts with) would be one in which 'semantic relations between expressions can be traced to (or be read off) their syntactic relations as established in the course of a derivation through the computational system.' More generally, Jackendoff's separation of syntax and semantics into syntax and 'conceptual structure,' two completely different generative systems, 'burden[s] us with an independent 'level of thought' (termed 'conceptual structure') that we will now have to assume as simply given and that nothing in the workings of the computational system underlying syntax will explain' – an extreme autonomy Chomsky never proposed.[157] SMT may be too strong, however – even in physics, which has reached levels of explanatory depth far greater than any other science, Max Planck still reminded his readers that 'we have no right to assume that any physical laws exist.'[158]

As far as the philosophy of language in concerned, however, Jackendoff clearly (in fact, emphatically) sides with the internalists. Opposing this tradition, externalist philosophers and linguists such as David Lewis and Robert Stalnaker take language as mapping onto 'possible worlds,' with the formal semantics of the logician Richard Montague, Barbara Partee and Barbara Abbott (for whom 'linguistic expressions refer to things out in the world') replacing the 'world' with 'model.'[159] This is 'a set-theoretic construct that enables the theory to be completely formalized. In principle formal semantics is neutral about its metaphysics. But on the whole "language" is taken to retain its common-sense status as something "out there in the world."'[160] The traditional study of metaphysics, then, is largely the linguistic study of semantic entities like objects, events and properties (though, at the same time, the concept of 'metaphysics' is synonymous with science, since it asks what the world is made of). This rigorous linguistic and conceptual analysis of traditional problems is what Locke identified as the concern of the philosopher, who is 'to be employed as an under-labourer in clearing the ground a little, and removing some of the rubbish that lies in the way to knowledge.'[161] Philosophy should be the mother of the sciences, as John Austin put it, setting up problems in such a way that sensible and meaningful empirical studies can then take over. Metaphysical questions of what the world is made of are today synonymous with naturalistic questions, leading Derek Gjersten to define metaphysics simply as 'the description of the heavens and the earth,' and for Aristotle to call it an 'awesome science' (indeed, his musings on declarative sentences in *On Interpretation* can also be seen as a precursor to Gricean pragmatics).[162] Made famous by a *Beyond the Fringe* sketch with Alan Bennett and Jonathan Miller, F. H. Bradley's belief was that the man who denies the existence of metaphysics is merely a metaphysician with a rival theory of his own. Cleaning up the 'wreckage sites' (Galen Strawson) of terms like 'consciousness,' 'matter,' 'capitalism' and 'democracy' still remains a primary concern of philosophers, popular science writers and political activists. Their task was set by Hertz's *Principles of Mechanics*:

> We have accumulated around the terms "force" and "electricity" more relations than can be completely reconciled amongst themselves. We have an obscure feeling of this and want to have things cleared up. Our confused wish finds expression in the confused question as to the nature of force and electricity. But the answer we want it not really an answer to this question. It is not by finding out more and fresh relations and connections that it can be answered; but by removing their number. When these painful contradictions are removed, the question as to the nature of force will not have been answered; but our mind, no longer vexed, will cease to ask illegitimate questions.[163]

The above terms also straddle the thin line between technicality and ordinariness (David Cameron's scrupulously misleading use of 'democracy' is a case in point). Joyce seems to had similar concepts in mind when writing *Ulysses*: 'I fear those big words, Stephen said, which make us so unhappy.'[164] Huxley exposed similar pretentions: 'We cover our anterior nakedness with some philosophy – Christian, Marxian, Freudo-Physicalist – but abaft we remain uncovered, at the mercy of all the winds of circumstance.'[165] Most of the problems of science and philosophy, then, simply revolve around finding out what the right questions are. Nagel's question 'What is it like to be a bat?' has no possible, imaginable answers, and so it's not a sensible question (it's rather a pseudoproblem, or a 'scheinstreiten' as Wittgenstein would have said).[166] As already pointed out, just because a sentence happens to have the structure of an interrogative expression it doesn't automatically get a free pass into 'questionhood.'

All of the above confusion is put in some perspective by Colin McGinn: 'The current state of the philosophy of mind, from my point of view, is just a reflection of one evolutionary time-slice of a particular bipedal species on a particular humid planet at this fleeting moment of cosmic history – as is everything else about the human animal. There is more ignorance in it than knowledge.'[167] In a recent essay on biolinguistics, Chomsky and Berwick survey other notable time-slices in the philosophy of language:

> For human language and thought, it seems, there is no *reference* relation in the sense of Frege, Peirce, Tarski, Quine, and contemporary philosophy of language and mind. What we understand to be a river, a person, a tree, water, and so on, consistently turns out to be a creation of what seventeenth-century investigators called the human "cognoscitive powers," which provide us with rich means to refer to the outside world from intricate perspectives. As the influential neo-Platonist Ralph Cudworth put the matter, it is only by means of the "inward ideas" produced by its "innate cognoscitive power" that the mind is able to "know and understand all external individual things," articulating ideas that influenced Kant. The objects of thought constructed by the cognoscitive powers cannot be reduced to a "peculiar nature belonging" to the thing we are talking about, as David Hume summarized a century of inquiry.[168]

It's consequently important to repeat that disciplines don't exist 'out there' in the world; rather, we invent them. The wealth of cross-disciplinary studies which relates the fields of literature, sociology, psychology and history (amongst others) preclude any 'barriers' that might separate inquiry of certain aspects of the world into definitive locations. The relevance for biolinguistics and philosophy is sketched by Hinzen: 'Organisms exhibit distinct features that we loosely classify as 'chemical', 'electromagnetic', 'mental'. We use these terms to indicate domains of inquiry, but no because we know what the mental or the chemical of its nature it. We do not use mental vocabulary to describe what we do because we know what mental properties *are*, and find them instantiated in some organism.'[169]

Many psychologists also subscribe to internalist interpretations of computational theories of vision, such as the neuroscientist David Marr's. Jacob comments: 'According to the internalist interpretation, computational-representational (C-R) theories of visual perception posit various levels of representation related by computational operations.'[170] Chomsky adds that, 'in the determination of structure from motion, it is immaterial whether the external event is successive arrays of flashes on a tachistoscope that yield the visual experience of a cube rotating in space, or an actual rotating cube, or stimulation of the retina, or optic nerve, or visual cortex.'[171]

To return to the question of concepts, if we use Chomsky's example of painting a house brown, it's understood, *a priori*, that it is the exterior surface of the house that is painted, not the interior (reflecting in part the CONTAINER conceptual metaphor). This particular way we look at the world is shaped by specific linguistic functions interacting with the visual system. This is because 'the semantic features impose an analysis in terms of specific properties of intended design and use, a designated exterior, and indeed far more intricacy. ... The exterior-interior dimension has a marked and unmarked option; if neither is indicated, the exterior is understood.'[172] The same also goes for any 'container' word, like 'box' or 'helicopter.' If I then say that somebody is near my house, we understand that the person is outside the house and not inside, despite both locations being equidistant from the house if they both stood, say, three metres from the surface. The phrase 'My empty house' refers exclusively to the interior, with 'My house is gone' suggesting the abstract, non-concrete nature of the object.[173]

James McGilvray elaborates: 'If in a cave and looking at its side, one is looking at the side of the cave, not a side of the mountain it is in. To look outside the cave at a mirror that reflects the outer surface of the mountain is, however, to look at the mountain's side.'[174] To show how this is truly an innate principle of the mind and not crafted through experience, if I were to say 'I painted the spherical cube brown,' it's understood that the exterior surface is painted and not the interior surface, despite the non-existence of this impossible, invented entity.[175] The spherical cube also has a designated interior and exterior, as the person standing near the spherical cube is positioned outside it, not inside. Surveying other evidence, Chomsky concludes that 'Over a large range of cases, we think of an object somehow as its exterior surface, almost like a geometrical shape.'[176] In *The Stuff of Thought*, Pinker explores further geometrical peculiarities:

> In the simplest case we can think like geometers and conceive of a point as having zero dimensions, a line or curve as having one, a surface as having two, and a volume as having three. But we can also conceive of more complex shapes by combining and ranking the dimensions. An object is thought of as having one or more primary dimensions, which are what really count in reasoning about it, together with one or more secondary dimensions. A *road*, a *river*, or a *ribbon* is conceptualized as an unbounded line (its length, which serves as its single primary dimension) fattened out by a bounded line (its width, which

serves as a secondary dimension), resulting in a surface. A *layer* or a *slab* has two primary dimensions, defining a surface, and a bounded secondary dimension, its thickness. A *tube* or a *beam* has a single primary dimension, its length, and two secondary dimensions, plumping out its cross-section.[177]

These form what Locke would have called 'internal conceptions,' and what Chomsky today calls 'rich perspectives for interpreting the mind-independent world.'[178] Ralph Cudworth expressed a similar internalist view in the seventeenth century: For him, an object is not 'stamped or impressed upon the soul from without, but upon occasion of the sensible idea excited and exerted from the inward active and comprehensive power of the intellect itself.'[179] It follows that these principles function the same way whether there's a mountain to see or a London to visit or a house to paint or whether there's no world at all: a lexical item is 'placed in an abstract space ... with its own organization that is only marginally influenced by the spatial relations in the real world.'[180] This is something Jackendoff often stresses:

> Now suppose there were a concept that had *only* inferential features. This then would be an abstract concept such as those expressed by *the value of my watch* and *the meaning of this word*. From the point of view of language, abstract cannot be distinguished from concrete objects: all that is necessary for reference is an indexical feature. The interfaces to language cannot 'see' whether a concept's descriptive features are perceptual or inferential. Thus as far as language is concerned, reference proceeds in the same fashion whether or not there is a conceptualized physical object out there in the world.[181]

Hinzen notes of internalism that we can conceptualise things as 'objects, relations, substances, and events, but from a physical point of view, everything out there may as well be a wave.'[182] Even if the externalist disagrees that London is no longer London if it is rebuilt one hundred miles away (judgments are 'delicate,' as Chomsky puts it), the 'very disagreement shows that the facts "out there" leave our judgements in such cases *open*.'[183] London is not a 'place,' as we've shown, though it may be *at* a place: indeed, as Hinzen notes, London could even become a 'virtual city' with no physical location at all, bearing no conceptual contradictions.

Similar principles of internalism apply to all aspects of biology, in particular neuroscience. Rodolfo Llinás explains that in neuroscience 'internalization' refers to the ability that the nervous system has to fracture external reality into sets of sensory messages (carried to the brain by millions of sensory nervous fibres) and to simulate such reality in brain reference frames. ... Thus a sensory input can be represented as a dynamic point vector in a system of sensory coordinates (the peripheral sensory systems) where the electrical activity in each fibre represents a component of such a vector. These sensory vectors represent, then, external objects (invariances) whose existence is independent of the system of sensory coordinates that encode them. In order to represent any such invariant in our brain, the sensory vector (in a multidimensional frequency space, since it is generated by impulses in many fibres) must be transformed into an internally meaningful vector. A truly fundamental problem becomes apparent here, since the reference frames intrinsic to the brain are different, of necessity, from the reference frames of the external sensory systems. Indeed, sensory responses arise from a direct interaction between the external world and a set of receptors which may respond to stimuli such as light, or sound or angular movement. Their messages in the form of nerve impulses must, therefore, be decoded by a set of internal neurons that do not share the external physical arrangement or the functional properties that receptor systems have in the periphery. This difference between the external and the internal reference frames relates to both the direction and the number of coordinate axes. ... These types of coordinate system-independent vector transformations are known in mathematics as tensorial transformations. [N]euronal networks actually implement tensorial transformation by means of their electrical activity and connectivity.[184]

In this sense, semantics is 'ultimately related to the problem of sensorimotor transformations. This term is used to address far more than a simple set of reflexes. Indeed, these transformations are viewed as utilizing sensory input to modify ongoing internal functional states. These internal functional states are then homomorphic with external reality. The sensory input feeds and modulates and internal state of intrinsic origin, that is, perception is a dream modulated by sensory input.'[185] Llinás' view is similar to Helmholtz's, who saw perceptions as the products of unconscious induction.[186] This tensorial approach to the nervous system could contribute to the science-forming capacity of human, being 'capable of formally describing aspects of mindness arising by evolutionary steps through the development of nerve networks. These nerve networks, by embedding the natural coordinates of the animal and then of the external world in relation to these natural coordinate systems, can develop an image of universals, particularly when helped by such tools as language and mathematics.'[187] Adopting a similar approach to biolinguistics, Hinzen has discussed with Higginbotham how contemporary philosophers

> think of the mind in relational, referential, and externalist terms. Realism is to see the mind's content as entirely reflecting the external world, rather than its own contents. Therefore, if you start emphasizing internal factors in the genesis of reference and truth, they think you are taking a step back from reality, as it were, and you become an anti-realist, or, even, a "Cartesian" philosopher. Now, early modern philosophers thought of all this in quite different terms. Realism and the objective reality of the external world was never the issue in Locke, for example, or even in Descartes, I would contend. What early modern philosophers and contemporary internalists in Noam's sense emphasize is internal structure in the mind, which underlies experience and enters into human intentional reference. But realism or denial is absolutely no issue in any of this. There is no connection between internalism in Noam's sense and an anti-realism or idealism, and that's in part because the relational conception of the mind is not endorsed in the first place.[188]

These and other insights make nonsense of Stephen Barker's claim that realism 'is alive and well, and untainted by metaphysics.'[189] Since language is used not to talk about the world as it actually is but rather how it is as conceptualised by the individual, Ludlow's comments also bear significant weight:

> With this rejection of referential semantics also comes a rejection of any attempt to use the semantics of natural language to gain insights into ontology. It is no good to argue from the structure of language to the existence of events, or plural objects, or times, etc. As Chomsky has argued, there are a number of constructions where the structure of language and the structure of the external world diverge. For example, some noun phrases intuitively have counterparts in the world (for example, the noun phrase "coats in the closet") while others do not ("flaws in the argument").[190]

In his *Lectures on Government and Binding*, Chomsky outlined the case himself:

> If I say "the flaw in the argument is obvious, but it escaped John's attention," I am not committed to the absurd view that among things in the world are flaws, one of them in the argument in question. Nevertheless, the NP *the flaw in the argument* behaves in all relevant respects in the manner of the truly referential expression the coat in the closet.[191]

The referentialist religion, then, commits us to impossible entities. The sentence 'The insect flew towards the light' might also appear harmless enough, but the words 'flew' and 'towards' both imply a purpose and intention. We also often ask whether computers can 'think' and whether mechanical arms can 'reach,' but not whether fire alarms can 'smell' or tape recorders 'hear.' This applies to the case when a tape recorder is left in a forest with no humans within hearing distance from a tree which falls to the ground next to it. Whether the noise the tape recorder 'hears' is a 'sound' depends on whether we ask a physicist or a semanticist. Moreover, the lower order of attention payed to smells compares poorly to the primacy of sight and thought. Our levels of sensory priority are therefore reflected in the properties we give to objects. McGilvray comments:

> Human concerns and intentions play an important role in our views of the identities of things. To burn a house is to destroy it. To dismantle a house, take the pieces somewhere else, and rebuild is to rebuild that house. To replace parts of a house (bricks, boards, beams) one-by-one and take the replaced parts elsewhere and build a house is not to rebuild the original house. ... So virtually any natural lexical language item (*not* scientific term) is a rich source of fine-grained distinctions that can be used by a person *because* they are a part of that person's linguistic *knowledge*, represented in lexical features.[192]

As Wittgenstein recognised, many of 'the aspects of things that are most important for us are hidden because of their simplicity and familiarity.'[193] Pierre Jacob comments on Chomsky's example of the sentence 'The book that he is planning will weigh at least five pounds.' In this sentence, 'the pronoun 'that' corefers with 'the book'. The pronoun refers to some content and its antecedent refers to a physical object that is the vehicle of the content referred to by the pronoun. In such cases, the anaphoric dependency of the pronoun on its antecedent is preserved in spite of the shift from the concrete to the abstract entity being referred to.'[194] 'Speakers of natural languages', he concludes, 'have subtle and intricate intuitions about the relations between an anaphoric pronoun and its antecedent.'[195] The word 'I' is also a 'flexible' term for Galen Strawson, since 'it can sometimes be taken to refer both to the self and to the whole human being, indifferently. Our thought (our semantic intention) is often unspecific as between the two; it's all pretty relaxed.'[196] On simultaneously abstract and concrete objects, Elbourne comments: 'We still have to explain how it is that we can say something apparently straightforward and true, like [abstract and concrete books], but using self-contradictory concepts. I am not aware of any work on this.'[197]

Jackendoff has also explored our concept of an 'end' and its various spatio-temporal manifestations, noticing intricate and subtle links between, say, the end of a speech and the end of a table.[198] The end of a line (projected onto something 'out there,' maybe what I would call a club but Sapir would call a pole) serves as the end of a boundary of a 1-dimensional object. The end of the pencil would be some arbitrary length (or the 'epsilon,' for Jackendoff). If I cut a centimetre off the fifteen-centimetre pencil, I can claim to have cut off the end of the pencil just as persuasively as my neighbour who cuts off two centimetres. Similar principles apply to temporal events, with the end of a speech also being arbitrary (based partly on individual interests). The end of a book could be interpreted as either its abstract temporal end (maybe the final chapter) or its concrete spatial end (maybe a perimeter spanning three centimetres around the cover). In his study of auditory events, Stephen Handel also points out that 'The perceptual world is one of events with defined beginnings and endings ... An event becomes defined by its temporal boundary. But this impression is not due to the structure of the acoustic wave; the beginning and ending often are not physically marked by actual silent intervals.'[199] Like 'ends,' motion can also be used metaphorically, as in 'He went from Liverpool to Nottingham' and 'He went from sick to well.'[200]

Diverging slightly, Jackendoff adds that

> it has often been noticed that the substance-object (mass-count) distinction in nouns (*water* vs. *bowl*) parallels the process-event distinction in verbs (*run* vs. *arrive*), and that there are often strong interactions between the two distinctions in the meanings of sentences. The parallelism motivates a semantic feature [±*bounded*] that contributes to both domains: objects and events have inherent boundaries, substances and processes to not. ... [I]t is in part this common distinction that allows us to say that both a table (object) and a speech (event) can have an *end*. But there is no perceptual similarity between the end of a table and the end of a speech.[201]

The principles of the mind outlined in the previous paragraphs have been thought to show that lexical items are used to point to particular regions of the conceptual-intentional interface. What's often been called the 'science-forming faculty' could potentially employ in its workings the sounds of language (phonemes), whilst not relying on the language centres of the brain to construct knowledge of the world. This could well be that case, but it's a factual claim based on the actual architecture of the mind – something we know fairly little about. As we saw at the beginning of this book during our study of the conceptual basis of the I/E-language distinction, a general expression like 'I went to the park' 'is accessed by performance systems that interpret it, on the receptive side, and articulate it while typically using it for one or another speech act, on the productive side. How is that done? The articulatory-perceptual aspects have been intensively studied, but these matters are still poorly understood. At the conceptual-intentional interface the problems are even more obscure, and may well fall beyond human naturalistic inquiry in crucial respects.'[202] Wittgenstein thought the words of a language were tools whose function is determined by the role they play in different 'language games' (they way we use language under specific circumstances), and in this respect his ideas seem similar to Chomsky's, as McGilvray notes:

> Chomsky agrees with Wittgenstein that meanings are uses/functions or applications of words (their role in truth-telling, to mention one popular option), there is no *theory* of meaning but only commonsense *descriptions* of use. Putting this in a Chomskyan-Cartesian way, words' functions are the uses to which they are put by free agents, uses that might be 'incited and inclined' [Chomsky, *Perspectives on Power*] by circumstance but are certainly not determined by them, whether internally or externally. Chomsky differs from Wittgenstein in other respects, however. Where Chomsky thinks that words – or at least, the sounds and meanings of words – are natural mental entities that enable us to deal with problems in a flexible way, not artifacts, Wittgenstein seems to have thought that people *manufacture* words and language to meet their needs [sic]. Furthermore – and most important – Chomsky thinks that the fact that words are not the products of human effort is precisely what makes it possible for human to be the creative creatures they are even at a very early age. Other rationalists made this point, as did Humboldt. Language – or the "ideas" or concepts of language – have to be innate and part of our nature if we are to manage to so quickly and in such a sophisticated manner deal with the world, play social roles, understand others' motivations, and create paintings and poems.[203]

This creative aspect of language is the core of Cartesian philosophy. Humans are 'incited and inclined,' for instance, to freely apply truth judgements to virtually everything: 'This *domain-generality* of truth it perhaps surprising given the high degree of specialization and "modularity" that we observe in cognition, action, and brain anatomy at large.'[204] Because 'Descartes' problem' of the source of linguistic creativity may lie beyond the scope of naturalistic inquiry, language *use* may have to permanently consigned to philosophy departments. It's even been suggested that the number of thoughts we can produce at any given moment is far greater than the number of particles in the universe. Indeed, it may exceed the number of natural numbers.[205] Descartes' view of human free will was close to A. W. Schlegel's, who thought animals live in a world of 'states of affairs' (Zustande) and not 'objects' (Gegenstande), unlike humans.[206]

But the idea that 'sounds' simply refer to 'objects' of 'things' in the world is far too simplistic, and has been the basis of many misleading and confusing debates in the philosophy of language since the 1950s. A coherent notion of things or objects in the world seems to most people to be taken for granted. Yet, in the same vein of the issues raised before over the mind-body problem, the fact of the matter is that a 'thing' is just an obscure a notion as 'sound' or 'word.' 'Things' like 'London' are usually understood to be external to the mind – something a physicist could examine. There also many concepts which lack lexical representations. Jackendoff has called these 'sniglets': To use one of his examples, the strips of grass which grow between the pavement and the curb have no name, but exist as concepts nonetheless.[207] This appears to be because 'discriminations along perceptual dimensions surpasses identification ... our ability to judge whether two or more stimuli are the same or different surpasses our ability to type-identify them.'[208] Here is Chomsky again on similar peculiarities in our conceptual-intentional interface:

> We do not regard a herd of cattle as a physical object, but rather as a collection, though there would be no logical incoherence in the notion of a scattered object, as Quine, Goodman, and others have made clear. But even spatiotemporal contiguity does not suffice as a general condition. One wing of an airplane is an object, but its left half, though equally continuous, is not. ... Furthermore, scattered entities can be taken to be single physical objects under some conditions: consider a picket fence with breaks, or a Calder mobile. The latter is a 'thing,' whereas a collection of leaves on a tree is not. The reason, apparently, is that the mobile is created by an act of human will. If this is correct, then beliefs about human will and action and intention play a crucial role in determining even the most simple and elementary of concepts.[209]

These questions were touched on by the seventeenth century Neo-Platonists, Hobbes, various theologians, as well as the 'empiricists' like Locke and Hume. Later they were briefly explored by the Gestalt psychologists of the 1920s and '30s (Wetheimer, Köhler, Koffka and others), but aside from some important work in the 1980s and '90s by

Jerry Fodor and his colleagues questions of 'lexical semantics' and concepts have remained largely ignored, with the issues surrounding the importance of intentions in determining 'objecthood' remaining virtually untouched even to this day (though we should also avoid Fodor's tactic, who 'often seems satisfied to explain a concept by typing the word for it in different cases and fonts'[210]). Jackendoff notes in his *Foundations of Language* how, 'On the one hand, [Chomsky] has argued vigorously for an internalist (here called 'conceptualist') approach to meaning, for instance in the papers collected in [*New Horizons in the Study of Language and Mind*]. On the other hand, aside from presenting a few telling examples, he has never attempted to develop a systematic internalist approach.'[211] Ignoring the fact that Chomsky is one of the few to even bother presenting internalist alternatives to mainstream philosophy and linguistics, attempting to explore a part of the world as obscure and lacking in comparative examples as the human conceptual structure has to be anything but 'systematic' at the current stage of limited understanding. Instead, groping towards the light in the form of lexical analysis (of the kind outlined above) and syntactic theory seems to be the surest way to reveal important and surprising insights into the intricate relations and interdependencies between semantic entities like events, quantifiers, variables and geometric surfaces – the closest we may be likely to come to a theory of meaning.

On the related question of why a noun phrase (like 'many people') alone cannot be a sentence, 'there may well be no principled explanation for it in terms of bare output conditions: the conceptual-intentional systems putting linguistic structures represented at the semantic interface to communicate use do not seem to necessitate such a distinction.'[212] One particular, internalist explanation has been given by Carstairs-McCarthy, who claims that 'the distinction derives from the exaptation [or 'characters, evolved for other usages (or for no function at all), and later "co-opted" for their current role ... Adaptations have functions; exaptations have effects'[213]] of an evolutionary prior physiological trait, the lowering of the human vocal tract. This allows vocalizations with a syllabic structure, which is then "exapted" for syntactic structure, which will mirror the structure of the syllable.'[214]

Adding to the mayhem surrounding 'objects,' Ronald Langacker argued in 1998 that the syntactic category 'noun' is equivalent to the semantic category 'thing,' where 'a "thing" is ... defined as any product of grouping and reification.'[215] But as Jackendoff rightly counters, this misses the crucial point that many nouns are not 'things' at all, like 'earthquakes and concerts and wars, values and weights and costs, famines and droughts, redness and fairness, days and millennia, functions and purposes, craftsmanship, perfection, enjoyment, and finesse.'[216] Many objects are also not 'physical' (in the common-sense use of the term, meaning perceptible and usually tangible), such as fictional characters like Jay Gatsby.

Writing on mental pathology, Henry Maudsley outlined how the categorising urge implied by Chomsky is constantly at work in all of us, since we have

> a sufficiently strong propensity not only to make divisions in knowledge where there are none in nature, and then to impose the divisions on nature, making the reality thus conformable to the idea, but to go further, and to convert the generalisations made from observation into positive entities, permitting for the future these artificial creations to tyrannise over the understanding.[217]

This biolinguistic perspective follows an older though much forgotten tradition in biological inquiry, stemming from the seventeenth century and Aristotelian thought. An archetypal internalist position was given in his treatise *On Interpretation*: 'Spoken words are symbols of mental experiences.'[218] Aristotle also pointed out in his *Metaphysics* that the form of a house (its wood and stones) differs from its function. He realised that material constitution (or 'the facts of the universe' for Carnap) is only one factor we take into account when deciding what an object in the world is – a simple idea which has been the cause of countless paradoxes.[219] Take the Ship of Theseus paradox developed by Plutarch, in which the Athenians begin to replace all the oars and timber of Theseus' ship during a voyage until not a single piece of the original ship is left. Meanwhile, the discarded parts of the original have been reconstructed with complete accuracy to the original. The question then arises, 'which is the Ship of Theseus?' This question is often applied to cars and other objects, and it seems to follow from the above paragraphs that the ship ceases to qualify as Theseus' precisely when an individual alone decides (including Trigger from *Only Fools and Horses*, who claims to have owned the same hair comb for decades, despite the fact that he often replaces its handle and head). For Hume, ships and houses 'endure a *total* change, yet we still attribute identity to them, while their form, size, and substance are entirely alter'd.'[220] The Theseus and Sorites paradox are inherently graded – but it's the concepts of 'ship' and 'sand heap' which are graded, not the external world itself. There's consequently nothing in the world which could possibly satisfy our complex concepts of house, river, ship and so on (to put it in a Lockean fashion, our concepts of 'river' and 'person' do not have 'resemblances' in the world), and so meaning 'is back in the head.'[221] Questions of 'objecthood' and Kripkean essentialism should be cognitively, not metaphysically: H2O, after all, if not an 'English' word or concept, but an invented theoretical entity like 'c-command.'

Contrary to the referentialist religion, no matter how we look the world simply won't tell us whether an object is a 'ship' or not. Though a child is never told so by their parents, and though the OED doesn't say so, somehow they have the knowledge that if a cow is cut into two halves it is still *a cow*, but if it is cut into dozens of small pieces and put into a large bucket it becomes *cow* (or *cow meat*), not *a cow*.[222] Chomsky also points out that a pile of sticks lying on a forest floor are a 'thing' if left there as a signal by someone, but not if they appeared there as a result of a forest fire. As with Putnam and Kripke, Jackendoff's 'decomposionalist' attempt of decomposing concepts into primitives still does no tell us what the concept actually *is*: 'Whatever the analysis we provide of *cow*, *paint*, or *chair*, our knowledge of this analysis, as long as it does not *contain* the concepts to be explained, is logically consistent with our lack of knowledge of the concept analyzed.'[223] Favouring an atomist theory of concepts, Hinzen argues

> not only for the primitive status of lexical atoms but for their logical independence for their physical representations and their neurological or external correlates. This would mean that even though there can and should be a programme of finding the neural correlates of single concepts, the problem of why these correlations exist and how the correlates can be explanatory for the concepts will not be resolved by more empirical inquiry.[224]

Understanding more than simply the mechanisms which enter into the use or formation of language and concepts is necessary for a full account of the structure of semantics. Making a point similar to one made by Galen Strawson in 'Against Narrativity,' Hinzen adds that the mind generates an illusion about its own transparency, since we 'unavoidably think that we have "reasons for our actions"', that we "know which rules we follow", but we do not. We do not in general control or direct our thoughts, we just think. ... We talk of persons, cities, houses, and chairs, and all of these things are distinct in highly specific ways, but ultimately we cannot explain, in a non-circular fashion, what causes them, and makes them distinct.'[225]

Like London, and any other 'thing,' the Ship of Theseus is simply not an entity of the extra-mental world. The notion of 'reference,' much discussed in the philosophical literature, seems not to exist at all in natural language. Plutarch presents no paradox if we simply allow two different perspectives: perspectives formed from a relation between the mind and the external world. This is the most revealing (and, I think, legitimate) study of semantics, much ignored since the eighteenth century. As George Lakoff has warned, 'The accumulated weight of two thousand years of philosophy does not go away overnight. We have been educated to think in those terms.'[226]

Emphasising again the importance of individual concerns in semantic interpretation, it's useful to recall Leonard Talmy's observation in 1978 that the notion of a 'nearby' star is metrically different from a nearby piece of furniture.[227] Nearness is 'defined in terms of the normal distance expected among individuals of the category in question'; with the same process applying to any evaluative adjective such as 'excellent.'[228] Some uses of the word 'river,' then, refer to physically identifiable entities, but there exists no category of such entities which is identifiable even in principle – leading Hume to conclude that such constructions of the mind are 'fictitious.' For the Dutch physiologist Jacob Moleshott, 'Except in relation to the eye, into which it sends its rays, the tree *has no existence.*'[229] Chomsky has also outlined how 'The semantic properties of words are used to think and talk about the world in terms of the perspectives made available by the resources of the mind.'[230] Chomsky has consequently urged that we reverse Aristotle's dictum from 'language is sound with meaning' to 'language is meaning with sound.'

A central theme of Aristotle's philosophy, as distinct from Socrates' and Plato's, was that, 'As in human production and activity, for example in the building of a house or ship, the idea of the whole is always the first thing present as the end of the activity, and as this idea then, by the carrying out of the parts, *realises itself in matter*, nature must be supposed to proceed in the same way, because in his view this sequence of end and thing, of form and matter, is typical of all that exists. After man with his aims, the world of organisms is established.'[231] Aristotle's metaphysics, from this perspective, seems not so metaphysical after all, but rather more epistemological, almost anticipating Locke's distinction between primary and secondary qualities: 'The bronze of a statue, for example, is the matter; the idea of the work is the form; and from the union of the two results the actual statue.'[232] Along with 'form' and 'matter,' Aristotle's metaphysics introduced two other notions of existence; 'efficient cause' and 'the end' (or 'final cause'), virtually principles of folk psychology (for and object to be moved it requires a force, and so on). However Aristotle, unlike Locke, tried to describe the actual nature of 'houses' – a metaphysical, and hopeless, issue – rather than our concept 'house,' a cognitive issue. Hobbes decried such investigations, and 'there is a specially notable passage in the forty-sixth chapter [of the Leviathan], where he indicates the confusion of name and thing as the root of the evil. ... Aristotle has made "being" into a thing, just as though there were in the universe an actual object which could be designated by the term "being!"'[233]

To follow on from this, we regard 'things' as human creations but not natural objects. The Theseus' boat paradox is applied to cars and helicopters, but not to trees, mountains or even animals. For I-languages, there are no determinate mind-external objects picked out by elements of phonetic representations as 'their sounds'; rather, these internal phonetic objects provide information to yield or interpret mind-external entities in a variety of ways that depend on circumstance, expectations, and so on. We consequently have good reason to agree with Chomsky when he claims 'there is no word-thing relation of the Fregean variety, nor a more complex word-thing-person relation of the kind proposed by Charles Sanders Peirce in equally classic work in the foundations of semantics.'[234] There is also no 'referentialist' relation between a particular hand gesture (as in the case of sign language) and a 'thing' in the world. As Peter Strawson pointed out over half a century ago (and often ignored since), 'referring' is something people do, not what words do. For John MacNamara, the act of referring to an object is an innate 'primitive of cognitive psychology.'[235] What is actually learnt during the acquisition process is not the meaning of a word, but rather the conventions of symbol use. Donald Davidson also agreed with Strawson:

> The intention to be taken to mean what one wants to be taken to mean is, it seems to me, so clearly the only aim that is common to all verbal behavior that it is hard for me to see how anyone can deny it ... if it is true, it is important, for it provides a purpose which any speaker must have in speaking, and this constitutes a norm against which speakers and others can measure success of verbal behavior.[236]

In an internalist theory of meaning, Jackendoff writes, 'reference is taken to be at its foundation dependent on a language user – just as relativistic physics takes distances and times to be dependent on an observer's inertial frame.'[237] Even in conventional usage, the deictic pronoun 'that,' as in 'Look at that,' has descriptive content only insofar as it denotes something in a relatively distal position (as opposed to the proximal 'this'). Its semantics is purely referential. After the expression has been externalised, it is up to other sense modalities to detect what the speaker it referring to. Jackendoff makes the crucial point that a visual percept will be linked to the deictic 'that' 'through the interfaces between conceptual structure and the 'upper end' of the visual system. Thus language has indeed made contact with the outside world – but through the complex mediation of the visual system rather than through some mysterious mind-world relation of intentionality. Everything is scientifically kosher.'[238]

How the hearer follows the gesture of the speaker is another prerequisite cognitive capacity for language growth, apparently unique to humans. But the visual system is insufficient for identifying 'things':

> The retina is sensitive only to distinctions like "light of such-and-such color and intensity at such-and-such a location on retina" and "dark point in bright surround at such-and-such a location on retina." The retina's "ontology" contains no objects and no external location. Nor is the situation much better in the parts of the brain most directly fed by the retina: here we find things like local line and edge detectors in various orientations, all in retinotopic format – but still no objects, no external world. This is all the contact the brain has with the outside world; inboard from here it's all computation.[239]

Linguistic entities, then, are no longer to be thought of as 'symbols' or 'representations,' but at 'structures' composed of discrete combinatorial units (a proposal Chomsky has had a hard time persuading philosophers to adopt over the years). These issues are found not just in language, but in various sensory and cognitive modalities (with David Marr's *Vision* laying the foundations for a computational account of the visual system), but the 'internalist' framework of inquiry derives from the philosophy of language.

The use of deixis is also not contingent on the existence of external objects, since we can say things like 'There is absolutely nothing right here.' Deixis can also fall prey to overextensions, as when Joey Mousepad, member of the Robot Mafia in the cartoon *Futurama*, flies through space overlooking Earth with his partners in crime. Being chased through the galaxy by the cops for robbing over $8 million ('That's a lot of De Niros,' mob boss Donbot comments), the gangsters try and think of a safe place to dump the money. Looking down on Earth from their spaceship, Joey points and says 'How about that dumpster down there?' 'You mean the one in that alley off 22nd Street?' asks his partner, Clamps. 'Yeah, what do you think I'm pointing at?' replies Joey.[240] Even the actual act of referring 'points' not to a presupposed object, but rather to some imposed geometrical surface: The directive 'Put your hat there' refers not to, say, a chair, but rather the abstract 'place' on top of it.

Initially observed by Charles Fillmore in 1982, the verb 'climb' also possesses intricate features which require a biolinguistic analysis.[241] In order for something to 'climb' it needs to tick one of two boxes (or both): It needs to be moving upwards, or it needs to exercise the human-specific motion of clambering (arms moving, hands grasping etc.). If it ticks neither of these boxes, it's not 'climbing,' whatever else it may be doing. If someone is clambering down a cliff, then, he's climbing. If an airplane is flying upwards through the clouds, it's also climbing. But if I say 'The airplane climbed down through the clouds,' it sounds wrong, since it's neither clambering nor moving upwards. Jackendoff introduced the non-Boolean operator 'Ψ-or' (where 'Ψ' stands for 'psychological') for meanings which are constructed 'by means of conditions connected with logical operators,' with 'climb' therefore meaning 'rise Ψ-or clamber.'[242] This 'check-list' theory of semantics claims, then, that each object has a certain threshold that its individual attributes ('feathery,' 'two legs,' 'can fly,' etc.) need to reach in order for it to qualify as that particular object ('bird').

Cluster concepts are also characteristically formed from a non-Boolean interaction of conditions. There are some suggestions that even prepositions may display these attributes, as explored in Simon Garrod, Gillian Ferrier and Siobhan Campbell's 1999 paper on spatial prepositions.[243] They explained how 'in' has both a functional and geometric component. Jackendoff elaborates that 'The geometric component stipulates that if X is in Y, X must be geometrically within the interior of the region subtended by Y. The functional component is 'containment': roughly, X is not attached to Y, but if one moves Y, X is physically forced to move along with it.'[244] The expression 'The ship is in the bottle' follows these stipulations, but 'The knife is in the cheese' does not, even though it still 'works.' It also works even if only a relatively short length of the knife's blade has penetrated the surface of the cheese. The knife and the cheese satisfy the functional condition, but 'when the objects being related are such that the functional condition cannot be satisfied, such partial geometric inclusion is far less acceptable' – a square partly overlapping a circle is not 'in' it.[245] Jackendoff calls this system of conditions a 'preference rule system.'[246] Since these concepts are a combination of conditions, combined differently and uniquely, traditional philosophical assumptions (including the Tarskian variety) must be abandoned.

Our intuitions are also sensitive to function and use, not just geometrical arrangement: 'A light bulb is considered to be *in* a socket when its base has been inserted, since that allows it to be illuminated, but a person is not *in* a car if only his arm extends in through a window, since that doesn't allow the car to move him or even shelter him.' We also compromise between functionality and geometry in determining the spatial relation of objects, a 'tube of toothpaste is judged to be *above* a toothbrush not when it is directly over its center of mass but when it is closer to the brushes.'[247]

Such a complex interaction of factors determining the meaning of a word 'can be plausibly instantiated in the brain, where the firing of a neuron is normally not a rigid function of other neural firings (like a logical conjunction), but rather a composite of many excitatory and inhibitory synapses of varying strengths. Thus cluster concepts, even though unusual in a logical framework, are quite natural in a psychological framework.'[248] Contextual factors can also crucially result in an entity 'being individuated by occasions in which it takes part in the activity: American Airlines may have carried three million *passengers* but only one million different *people*, thanks to a lot of frequent flyers.'[249] Though she rigorously defends the incoherent E-language notion of 'public language,' Ruth Millikan's concept of 'proper function' develops the above points quite usefully. Having a proper function, she writes, 'is a matter of having been 'designated to' or of being 'supposed to' (impersonal) perform a certain function.'[250] Mail can still be mail, then, even if fails to arrive at its intended destination (this is not a necessary condition for mail to still be mail, but still an important one). The 'thing' may never actually perform its function and still remain a 'thing,' since the mail may be left in the office before being accidentally shredded. As Jackendoff qualifies, 'artefacts' like pens (concrete) and myths and beliefs (abstract) also have proper functions. So too do 'parts,' like the back of a chair and bodily organs. Humans and some animals (like cats and sheep) also have 'occupations,' 'a special type of proper function, relativized to humans and domestic animals.'[251]

The biologist Edward O. Wilson once called for 'a deeper and more courageous examination of human nature that combines the findings of biology with those of the social sciences.'[252] As a way of heading towards this goal, important studies by Deborah Keleman have shown that children have an innate tendency to assign purpose and agency to natural events.[253] Answering questions about what certain natural environments are 'for,' young children mostly reply that, for instance, clouds are for raining, sharp rocks exist so animals can scratch their backs on them, the water in lakes is cool so that fish don't get too overheated and exhausted, and so on. The disposition is so powerful that we would not be mistaken to conclude that humans are natural-born creationists, 'and many never grow out of it.'[254] The very fact that children so easily grasp the meaning of words like 'persuade' strongly suggests an innate internal mechanism that assigns agency and causality in unique, instinctual ways. 'All events in nature', writes Wartofsky, 'are conceived under the form of willed events, and the doers of these events enact them as *dramatis personae*.'[255] It's of little surprise, then, that Jackendoff points out that children literally ascribe proper functions to things. 'Driver' is conceptualized according to the object or agent it relates to: 'If a person, it can denote a person who is driving or whose occupation is driving; if an artifact, it denotes something whose proper function is driving (e.g. a gold club or a driving wheel on a locomotive).'[256]

All of the above saves us from falling into the trap of simply concluding, like a good Wittgensteinian, that 'language is use' and leaving it at that. But Chomsky gives good reasons not to resort such behaviour. It's been said by Jeremy Bentham that the world is divided into two types of people: those who divide the world into two types of people, and those who don't. Chomsky certainly does, with his internalist critiques of externalism, largely ignored in the philosophical community since their publication over a decade ago: 'Note that in dealing with both the phonetic and semantic aspects of language, the internalist approach adopts as a matter of course a certain form of "externalism", but one too weak to be of any interest: that observation of usage plays a role in establishing some properties of an expression, its sound and meaning. To be of any significance, externalism must go well beyond that truism.'[257]

Echoing, Austin, Jackendoff reminds us that 'making assertions that can be true or false is only one of many things we can do with language. Theorists' concentration on truth value, which seems to go all the way back to Plato, blinds us to the full vivid range of possibility.'[258] Talk of 'truth-conditions' on contemporary philosophy of language is almost like talking about 'interrogative-conditions.' Though many declaratives are not intended as assertions ('We all live in a yellow submarine'), their 'truth-value' is not an issue to either the speaker or listener. To use one of Jackendoff's favourite examples, the peculiar nature of truth is exhibited by examining fictional entities: It's true that Sherlock Holmes was English, but false that he existed. The intended fictional/non-fictional status of a lexical item is irrelevant in terms of grammatical processes, hence the reason why 'we can "build" possible worlds as much as we can depict the actual one, and in either case proceed with equal ease, using, from a linguistic point of view, the exact same mechanisms and combinatorial principles. In sum, language is insensitive to the difference that existence makes and it leaves ontology behind.'[259] We should recall Parmenides' claim that truth is not tensed, adding to it Hinzen's adage that 'Truth is for sentences what existence is for noun phrases.'[260] Leaving ontology even further behind, he points out that no language 'has any morphological marker existence the way that it may have markers of animacy, tense, or social status.'[261] With no 'existence' marker, 'King Lear is mad' is just as acceptable as 'William Hague is mad.' It is strange, then, that countless philosophers have analysed fictional names as descriptions, but not real names, with no syntactic reasons being given behind such a distinction.

In his later writings Rudolf Carnap asks us to consider asking a 'man on the street' whether the following sentence is true, given the non-existence of unicorns: 'A unicorn is a thing similar to a horse, but having only one horn in the middle of the forehead.'[262] But as Carnap failed to point out, syntax does not discriminate between fact and fiction, and so there's no reason to think that semantics (or phonology) would either. And unlike Kripke and Searle, who claim that a name's meaning is its mind-external 'reference,' biolinguistics counters this philosophical stipulation by arguing that 'referential features of names are centrally a consequences of their syntactic form, and their conceptual content is not explained by their reference. Hence, even at the level of the lexicon syntax enters into the explanation of meaning.'[263] By Jackendoff postulating an extraneous 'conceptual structure,' he seems to find these concerns about the importance of syntax unimportant, even boring. He claims, for instance, that because some context-sensitive expressions like 'Don't!' (when spoken by someone watching another person about to jump off a tall building) lack a 'deleted sentence' embedded in them (e.g. 'Don't [jump off the building]!') there must be a separate 'conceptual structure' which counters the 'syntactocentric' model of biolinguistics. But some lexical items simply don't have edge features (also a requirement for objects to be merged) and combine with others, being interpreted solely as interjections (ideas inspired by X-Bar theory, in which lexical items are 'projected' up through the structure carrying with them 'features' which determine how and where they can combine).

But along with the true non-existence of Sherlock Holmes, Jackendoff does however identify similar quirks in the following sentences, focusing on the pronoun 'it':

> 1. There was a performance of *Harold in Italy* by Zukerman on Thursday. I heard *it*. *It* was fabulous.
> 2. Zukerman performed *Harold in Italy* on Thursday. I heard *it*. *It* was fabulous.

In '2,' 'I did not hear the truth-value of the proposition that Zukerman performed, and I did not think this truth-value was fabulous! Rather, just as in ['1'], I heard the *event*, and the *event* was fabulous. In ['1'], *it* corefers with *a performance* ..., and in ['2'], *it* corefers with the entire first sentence. Thus the sentence too must refer to the event. (Such evidence is rarely considered in the philosophical literature on reference, although it was a mainstay of the briefly fashionable theory of Situation Semantics [of Barwise and Perry's 1983 *Situations and Attitudes*]).'[264] Considering further eccentricities, Hinzen presents the sentence 'No head injury is too trivial to ignore,' which when spoken is understood to mean 'No matter how trivial a head injury is, it should not be ignored.' But the expression actually means 'No matter how trivial a head injury is, it should be ignored': 'What an expression would be 'sensibly', 'rationally', or 'usually' interpreted as saying need not be what it means. The system has its own mind, so to speak, working by its own idiosyncratic principles, apparently not 'rational' ones (in something like the sense of rational decision theory).'[265] Truth is therefore strictly a human concept, a property we ascribe to sentences and not independently given in the world. But we have no answer to the question 'What is truth' largely because we have no answer to 'What is a word?' In the first page of his *Slim Guide to Semantics*, Paul Elbourne points out that, 'Despite 2,400 years or so of trying, it is unclear that anyone has ever come up with an adequate definition of any word whatsoever, even the simplest.'[266]

It's been noted by Bruno Snell that the Greek language, by inventing the definite article ('the'), 'could take an attribute of an existing thing, expressed through an adjective – that is was 'beautiful', say – and turn it into an abstract noun by adding the definite article: so from beautiful (*kalos*) to 'the beautiful' (*to kalon*).'[267] The definite article is not the *cause*, though of the abstract/concrete distinction, but one of its potent mechanisms. 'The' also 'orients an idea as identifiable to the listener [and] has been responsible for numerous articles and books and is still and object of controversy.'[268] Though determiners and adjectives often have similar syntactic functions (they can both be positioned pronominally, as in 'the ship' and 'big ship'), adjectives – like adjuncts ('in the garden,' 'with a drink')[269] – can be recursively stacked ('big, heavy, sturdy ship') unlike determiners ('*a my that the ship' – where the asterisk denotes ungrammaticality).

Taking the sentence 'The enemy destroy the city,' which could be tensed to produce 'destroys' or 'destroyed,' Hinzen has examined the intricate function of the definite article. The verb phrase ('destroy the city')

> strictly comes to be *about* something only once it gets, in the process of building up structure, tense: this is the "deictic" function of Tense. On analogy, D [determiner, 'the'] would also comprise the "deictic element" in the nominal domains, where again NP ['the enemy'] itself would be purely predicative: *city*

does not refer to anything specifically, but *the city* (in a particular discourse) does. A philosophically intriguing simplification, Kantian in flavour, would be that T [tense] locates in time where D locates in space.[270]

So for something to be 'true,' Hinzen continues, we need to position our truth-judgement in time through tense:

> Introducing in the place of the notion of a judgement the metaphysical category of a "fact" for what sentences express, and saying that a "proposition" "denoted" by a tensed VP (or an IP [an extended projection of a verb which heads a sentence]) corresponds to a "fact", obscures this: a fact does not contain Tense, and neither does a proposition (though it is hard to tell, as that notion, as a purely technical one, will depend on how it is defined).[271]

We can, then, draw the following conclusion with Jackendoff that 'the form of these [truth] conditions is constrained by human psychology, not by logical necessity.'[272] There are thus no empirical 'theories' of propositions (except metaphysical and normative ones), there is only syntax: 'Propositions are strangely intermediate entities between what's in the head and what's out there in the physical world.'[273] Asking 'Is this sentence true?' is consequently meaningless (as is the question 'Is this particular action moral?'). We should instead be asking 'Is this sentence true for a particular person, and if so what are the syntactic operations and constraints which yield such truth-judgments?'

Equating language with communication, Dummett even wrote that truth is synonymous with the aim of an assertion, a view which has peculiar undertones of an adaptationist view of language.[274] James' similar instrumentalist view of truth held that 'The possession of true thoughts means everywhere the possession of invaluable instruments of action.'[275] Sentences do not have 'truth-conditions,' but rather 'truth-indications,' to use Chomsky's term.[276] Flying against a dominant assumption in philosophy, 'The child is no "representational medium", as there *is* nothing in its environment "of" which the principles of UG in its head could be "true."'[277] The prominence of vagueness (and its common response of contextualism) adds weight to the claim that 'truth-indication' is a fit term for natural language semantics.

Tensed sentences can be eternal in Quine's sense, having the same truth-value over time. Commenting on a concern raised by G. E. Moore, James Higginbotham has considered how it's possible to say the same thing twice using a simple past tense.[278] This is what Davidson called the 'fly in the ointment: the fact that the same sentence may at one time or in one mouth be true and at another time or in another mouth be false.'[279] When 'Caesar was murdered' is uttered twice, Higginbotham notes, 'I seem to say in the first utterance that he was murdered before that, and in the second that he was murdered before *that*.' This question

> arises whether tenses are taken as operators or as involving temporal points and predicates, since utterances of the same past-tense sentence at different times will have different truth-conditions. But if we integrate our understanding of context with the purely semantic exposition, the answer is clear: even if you can't, in a way, say the same thing twice by saying "Caesar was murdered", we know that if any utterance of "Caesar was murdered" is true then all subsequent ones are. And where we avow "Many people have said that Caesar was murdered", we count ourselves as speaking truly despite the difference between the truth conditions of any two non-simultaneous past utterances of "Caesar was murdered", because we know that the context links them so as to make them true or false together. We don't need identity; closeness of fit is enough.[280]

Consequently, Higginbotham takes this as 'another pointer toward the need for integration of contextual and linguistic material in the interpretation of natural languages. The larger lesson, both for logic and philosophy, is that we should be prepared to elaborate systems and conceptions of truth and of consequence that show context and language working together.'[281] Hans Reichenbach's analysis of the English tense system also stressed the importance of the 'reference time' an utterance refers to, along with the time of the event and the time of the utterance (this stress on reference as an individual action, and not a metaphysical relation between linguistic entities and external objects, lost its sway on the philosophy of language soon after Strawson's paper 'On Referring' a few years later).[282] Thoughts such as these exhibit 'intentional stability' which are not disrupted by external parameters: 'my water bottle' will remain intentionally intact even if I do not have a water bottle at 'this' particular moment but rather intend to buy one ('I will put my water bottle right here,' one might add), or if the object I consider to be a water bottle turns out on closer inspection to be a shoe.[283] Dennett makes the sound point that we should expect an inherent indeterminacy to exist between a natural kind and its reference, adopting a 'Darwinian' stance on meaning (though his meaning-theoretic story of horses and schmorses constrains semantics solely to another form of adaptation).[284] Hence 'water' remains conceptually coherent (or 'stable') whether its referent is H2O, XYZ, or anything else.

Like 'objects,' 'events' can also be divided intricately divided into achievements (instantaneous), accomplishments (process leading to a terminus) and processes (ongoing events which do not undergo a significant change of state).[285] Noting the relation between 'John,' 'runs,' and 'the running of John,' Hinzen postulates that

> it is not implausible that a verb is something like the "dynamification" of an object, and an object is the freezing of such a dynamic event, and that these two categories are fairly closely linked to a distinction at the level of the syntactic categories. ... [P]erhaps a verb *is* nothing other than its Theme or internal argument viewed dynamically. Thus, an event of *drinking a beer* is a *beer* viewed dynamically as a gradual reduction of its volume, until it disappears.[286]

The semanticist Paul Pietroski also proposes that predicate conjunction for a given event (e) is the semantic contribution of binary branching (or syntactic complementisation), which leads the *adjunction* structure [[*Jill killed* Bill] [*by* strangulation]] to have the following basic syntactic structure:

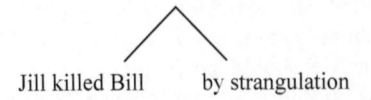

```
        /\
       /  \
Jill killed Bill   by strangulation
```

This means 'there was an event e of killing, of which two predicates hold: *Jill killed Bill* is true of it, **and** *by strangulation* is true of it.'[287] If the same method of interpretation is true of an *argument* structure such as

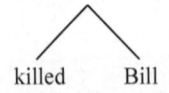

```
     /\
    /  \
 killed  Bill
```

Then we are left with the meaningless 'killed **and** Bill,' and we face two problems: where the conjunction comes from, and how 'killed and Bill' can make the same sense that *killed Bill* makes. Pietroski proposes that the conjunction is the semantic contribution of the syntax, which 'itself contributes the conjunctive aspect of meaning,' bearing the 'semantic load.'[288] As for the second problem, he concludes that 'the system, designed to blindly combine predicates, *turns* the argument NP *Bill* into a predicate, essentially the predicate "Bill is a THEME of e', or [THEME (Bill)] (e).'[289] Consequently, once we have a syntactic schema that we can extract from a transitive construction like Jill kills Bill, namely [α [Φs β]], translate brackets (branching) to conjunctions and add theta-roles, and allow ourselves an existential quantifier $ to bind the event-variable, we get the semantic form: (i) $e such that AGENT(α) is true of e and [e is a Φing and THEME(β) is true of e].[290]

This is another 'internalist direction for semantic interpretation,' Hinzen thinks, with syntactic form determining semantic form, helping to answer a core question in modern linguistics of how much meaning syntax can carve out.[291] Semantics is therefore distributed, in a sense, since 'the meaning that a lexical item has in the sentence in which it occurs comes about in stages during the derivation, rather than being there from the start, or coming about in one fell swoop in the end.'[292] Consequently, the external aspects of meaning, like historical events, fail to explain the meanings they are associated with. Spelling out this minimalist methodology to semantics, Bolender, Erdeniz and Kerimoğlu draw up the following principle:

> So far as possible, seek explanations of uniquely human concepts in terms of syntactic computations.[293]

The functionalist, however, believes 'thought' exists to 'represent' the world, with 'language' existing to communicate these thoughts. Syntax only arises after the pre-existing semantic 'contents' are mapped onto phonology and syntax. 'This viewpoint is the reason,' Hinzen points out, 'why even today, in classes on philosophy of language, the syntax (and phonology) of natural language is essentially ignored, as if it didn't belong to language.'[294] Additionally, we could take Kripke's claim that 'Gödel' remains 'Gödel' even if it were revealed he did not discover the Incompleteness Theorems of 1931. But the 'stability ("rigidity") of the object of reference across possible worlds that [Kripke] points to find explanation, not in this object, but in the concept that our predications involving it include.' This also applies to alleged 'natural kinds,' since 'nouns take kind-denotations only in particular syntactic contexts.'[295] When a name is in fact 'rigid' in its denotation, this is purely of the simplicity of its syntactic form.

There also words which have syntax and phonology but no semantics, as in the sentence 'It is hot in here,' where 'it' simply satisfies the imposed syntactic demands. Similar default words for syntactic constraints include 'of,' as in 'His fear of heights.' Other words have phonology and semantics but no syntax, as in 'ouch,' 'hello' and 'yes.' Verbs too sometimes specify some of the syntax and phonology, as in 'angry' (with/at) and 'count' (on), and so on.[296] Among the above network, Jackendoff argues, 'might be certain default principles that odd independent biases of 'constructional meaning' to subject and object position when the verb leaves the option open.'[297] He states two possible, and intriguing, examples:

> 1. Preferably: Subject corresponds to Agent; Object corresponds to Patient (entity affected by an action).
> 2. Preferably: Subject corresponds to Figure; Object corresponds to Ground.

Presenting examples made famous by Pinker, Jackendoff demonstrates these constraints using so-called 'spray-load' alterations:

> 1. We sprayed paint on the wall.
> We loaded furniture on the truck.
> 2. We sprayed the wall with paint.
> We loaded the truck with furniture.

The two pairs 'differ subtly in what is taken to be affected by the action, with somewhat greater emphasis on the point and the furniture in [2a] and on the wall and the truck in [2b]. This can be attributed to the effect of the Patient condition in [1a] on an otherwise ambiguous situation.'[298] As Beth Levin and Malka Hovav's discussion of argument structure alterations showed, 'sounds emission verbs can appear in motion verb frames if the sound emission can be associated directly with the action of moving (e.g. *The car squealed around the corner* but not *The car honked around the corner*).'[299] These and similar 'empirical results are part of what motivates the search for a new foundation [of meaning] ... The relation between the philosophy and the dirty work has to be a two-way street' – with much collaboration needed between philosophers, linguists, psychologists, neuroscientists and (with the emerging Evo-Devo program) biologists if any such internalist/rationalist perspective is to carry any weight.[300] Syntax of course strains externalisation, but, Jackendoff also asks, can intonation constrain syntax? He gives the following examples to suggest it can (where * indicates ungrammatical and ? indicates dubiously acceptable):

> 1. John bought a computer yesterday.
> 2. * John bought yesterday a computer.
> 3. ? [John bought several expensive pieces of hardware that he's been dreaming about for months] [yesterday].
> 4. [John bought yesterday] [several expensive pieces of hardware that he's been dreaming about for months].

As a plausible explanation for this phenomenon, Jackendoff explains:

> Normally the syntax of English is very insistent that the direct object precede any time adverbials, as seen in the contrast [1, 2]. But if the object is very long and the time adverbial short, the reverse order [4] is far more acceptable. The reason is that this permits the prosody to satisfy [certain] conditions much better: the IntPs [Intonational Phrases, namely 'the part of the sentence that falls under an intonation contour'] are much closer in length, with the longer at the end. So evidently the needs of prosody are forcing a non-optimal syntactic structure.[301]

Like Jackendoff (a fellow functionalist), Pinker in *How the Mind Works* takes the following naturalistic approach:

> Our minds evolved by natural selection to solve problems that were life and death matters to our ancestors, not to commune with correctness or to answer any question we are capable of asking. We cannot hold ten thousand words in short term memory. We cannot see in ultraviolet light. We cannot mentally rotate an object in the fourth dimension. And perhaps we cannot solve conundrums like free will and sentience.[302]

Problems like Braun and Saul's notorious puzzle of the 'resistance to substitution in simple sentences' also 'go the way of reference' (to use Chomsky's phrase[303]) when examined from an internalist/biolinguistic perspective.[304] Initially, their puzzle collapses both belief and meaning, since (1) seems true while (2) seems wrong, and should in turn lead to the falsehood of (1), given that Superman is Clark Kent:

1. Superman leaps more tall buildings than Clark Kent.
2. Superman leaps more tall buildings than Superman.

Hinzen counters, again emphasising the centrality of syntax, that 'there is a constraint operative in the human linguistic system according to which if there are two referential expressions in one clause, they are by default interpreted as not referring to the same thing (reference is obviative). The mechanism is a dump one: two nominals, two interpretations. ... [(2)] makes us look for different referents where the formatives involved indicate there is only one, which is what puzzles us.'[305] Small clauses and other special cases can override these constraints, such as in (3):

3. We called [him John]

Concluding his highly informative *Essay on Names and Truth*, Hinzen also suggests it is quite possible that 'part of where formal ontological notions such as part and whole come from is syntax.'[306] For him, integral relations are not metaphysical but syntactic notions (a subtle but revolutionary turn in the philosophy of mind), reversing a trend which has survived since Aristotle. As with Pietroski's event semantics, interpreting these notions as part of syntax leads to the conclusion that

> objects are wholes if *conceived* or conceptualized as objects-together-with a certain quantifiable structure of parts or aspects to which they may relate integrally: the aspects or parts are not cognitively speaking objects in their own rights, but have a specific inherent *connection* with the objects of which they are the aspects.[307]

Contrary to Quine's view that an understanding of substances and individuals depends on an acquired quantificational syntax (such as the mass-count distinction), Soja, Carey and Spelke have shown that during the earliest stages of language acquisition, children make use of conceptual categories of 'substance' and 'individual' virtually equivalent to those used by adults, forming our 'intuitive materials-science.'[308] Count nouns are unbounded and made of individuals, whereas the opposite applies to mass nouns, leading Pinker to suggest that 'our basic ideas about matter are not the concepts "mass" and "mass" but the mini-concepts "bounded" and "made up of individuals."'[309] Unlike pebbles and gravel, collective nouns like 'committee' are both bounded and made of individuals: but again, these are not metaphysical matters. 'Hair' and 'hairs' are respectively count and mass nouns, but a physicist would not ask with Richard Lederer 'why a man with hair on his head has more hair than a man with hairs on his head.'[310] The mass-count distinction is certainly no law of nature, since interpreting 'matter' as 'countable units or amorphous stuff' is a modifiable perspective: 'We can always look at a cup (count) but think about the plastic composing it (mass), or look at some ice cream (mass) and think of the shape it assumes, such as a scoop or a bar (count).' We can also 'defy a language's stipulation by mentally packaging the referents of mass nouns (*I'll have two beers*) or by grinding the referents of count nouns (*There was cat all over the driveway*).'[311] Frege's semantic theory is unsuitable for this, dealing as it did with countable 'things' only and not substances.

As with Jackendoff's 'perspectival perspective' and Chomsky's 'internalist' view of the mind, Pinker takes the mass-count distinction to be 'a cognitive lens or attitude by which the mind can construe almost anything as a bounded, countable item or as a boundariless, continuous medium.'[312] He also points to a unique type of mass noun, the mass hypernyms or superordinates like 'furniture' and 'mail,' which do not refer to a substance but, neither do they refer to individual objects (making the task of 'mail' remaining 'mail' even if it never reaches its destination somewhat easier). Pinker's adaptationist explanation for the emergence of the mass-count distinction (that it helped us... count!) could certainly be plausible, with counting being an impossible task without agreed and defined units. But the question of how brain cells create 'mass' concepts like 'goo' or 'water,' objects which have indefinite spatiotemporal extensions and have 'more' or 'less' of itself produced but cannot be counted, is 'so far beyond the scope of current scientific inquiry that evocations of Neo-Darwinism seem essentially meaningless.'[313] Pietroski and his colleagues have also found intriguing evidence that conceptualising a random display of objects on a screen as a substance (e.g. 'goo,' a mass noun) rather than discrete units (e.g. 'dots,' a count noun) aids in quick estimations of the quantity of objects. The origin and possible 'use' of this unconscious mental principle remains obscure.[314]

The mass-count notions of bounded/unbounded are not only also found in geometry (with a 'river' being an unbounded line, its length, with a bounded one, its width), in event semantics we also find 'bounded accomplishments (*draw a circle*) and unbounded activities (*jog*). Just as we met substance words that name homogenous aggregates (*mud*) and plurals that name aggregates made of individuals (*pebbles*), we now meet durative verbs that name a homogenous actions (like *slide*) and iterative verbs that name a series of actions (like *pound*, *beat*, and *rock*).'[315] Though lacking neuro-cognitive evidence, Boban Arsenijević speculates that there may exist

> parallels between places in cognitive maps and referents of nominal expressions in language on the one hand, and between paths in cognitive maps and eventualities in language. Both members of the former pair correspond to geometric points, and both members of the latter pair have linear structures. Moreover, both paths and eventualities include places and objects, respectively, as important defining elements in their structures.[316]

Language growth also influences spatial cognition:

> Some properties of the processing of cognitive maps, including the use of descriptive and geometric cues, undergo a drastic change around the age of six, which is also considered the

critical period during which the individual rounds her acquisition of grammar.[317]

In his short but fruitful philosophical career, Frank Ramsey also noted that to add the addendum 'is true' to a sentence is to say 'no more than the earth has the quality you think it has when you think it is round; i.e. that the earth is round.'[318] Placing questions of truth and reference in a biolinguistic setting, Hinzen makes the Kantian point that 'a proposition changes no more its content when we call it true than an object changes its nature when we judge it to exist.'[319] Speaking of the truth and the 'real' truth is not to speak of two different truths, either; it is merely to emphasise the extent to which one believes a certain truth to be true. Both in semantics and physics, 'real' is simply an honorific term (and, as J. L. Austin said, an 'excluder' which gets its meaning from what it is opposed to, as in 'real ice cream' versus 'artificial ice cream'); electrons are 'real' because they are 'physical' entities posited in our best explanatory theories. If the explanatory theory is proven false, the posited entities are abandoned too. Kuhn also felt that there is 'no theory-independent way to reconstruct phrases like "really true"; the notion of a match between the ontology of a theory and its "real" counterpart in nature now seems to me illusive in principle.'[320] John Collins' 'naturalistic reality principle' seems to be the most sensible approach to the matter, stipulating that 'At a given stage of inquiry, a category is taken to be (naturalistically) real iff it is either successfully targeted by naturalistic inquiry or essentially enters into the explanations of such inquiry.'[321] And so, 'after the development of the special theory of relativity, length, shape, simultaneity, etc. ceased to be "real" qua frame-dependent' in the same way that the classical categories of Earth, Air, Fire and Water are no longer 'real.'[322] We 'attribute "reality" to whatever is postulated in the best theory we can devise,' as Chomsky succinctly puts it.[323]

It's in this respect that the formal computations of generative grammar (thanks to the Principles and Parameters framework, the physical basis of which is yet to be determined), can be compared with Mendel's laws at the beginning of the twentieth century. In fact, with Higginbotham's 'panel of switches' metaphor for the Principles and Parameters framework, along with empirical evidence suggesting the apparent variation of languages is based on the setting of innate parameters, Baker believes these factors yield the linguistic equivalent of the periodic table of elements.[324] In their introduction to genetics, George Beadle and Muriel Beadle point out that 'There was no evidence for Mendel's hypothesis other than his computations and his wildly unconventional application of algebra to botany, which made it difficult for his listeners to understand that these computations *were* the evidence.'[325] In the early parts of the twentieth century, physics and chemistry were largely viewed from an instrumentalist perspective, with the likes of Ernst mach insisting in 1897 that atoms were only 'a mathematical model of facilitating the mental reproduction of facts.'[326] These views were influenced by a Machian outlook of science, which effectively claimed that if our sense organs can't pick it up, it doesn't exist (it isn't 'real'). William James also argued along similar grounds that the theories of chemists were not about the world, but rather useful tools for calculating the nature of specific chemical reactions. James argued that 'It is ... *as if* reality were made of ether, atoms and electrons, but we mustn't think so literally. The term "energy" doesn't even pretend to stand for anything "objective". It is only a way of measuring the surface of phenomena so as to string their changes on a simple formula.'[327] After the unification of chemistry with quantum physics, it became 'real'; perhaps similar things may occur between linguistics and neurobiology.

Just as Galileo's 'wildly unconventional' application of geometry to falling cannon balls and Chomsky's 'wildly unconventional' application of mathematics to linguistic data may be reduced or unified with neural accounts of language. But though he advocated idealizations Galileo failed to go as far as Newton, retreating from the non-mechanical forces of Renaissance naturalism: 'In this vein, Galileo ridiculed Kepler for his assumption of a magical "attracting force" acting between the moon and earthly water, and offered a (wrong) mechanical model instead.'[328] Advocating the need for such idealizations through theoretical linguistics, Hinzen argues that due to the 'completely different' nature and constituent features of the sound and meaning systems, it follows that 'meaning is not a matter of sound, and the language system itself, being a way of *relating* sound and meaning, does not reduce to *either*, being more abstract and 'central' to the mind than either of them.'[329] We should recall the premature judgment of Wolfgang Pauli that the postulation of electron spin was 'very clever, but of course it has nothing to do with reality.'[330] We also should recall how, while confessing that he lacked a cause for the 'absurd' properties of gravity and the inverse square law, Newton nevertheless concluded in his *Principia* that 'it is enough that gravity really exists.'[331] Newton's force 'really' exists for the same reason noun phrases and RNA 'really' exist – namely, that they are posited by our best explanatory theories and backed up by experimental observation. Maybe, too, 'it is enough' that free will and the creative aspect of language 'really exist,' and if their causes cannot be rendered comprehensible by our theories, too bad for our minds. Cries against this so-called 'mysterianist' position by the likes of Dennett seem to stem from the common assumption that because we can touch the world with our hands we can therefore grasp it with our minds. But 'Modern physics,' Hinzen points out, 'can make little sense of our common-sense notion of a "solid body"':

> Schrödinger gave up on the notion of the particle altogether, and quantum theory introduced further mysteries, from non-locality to the conclusion that given the role of the observer and his conscious decisions, the 'material' world is not wholly 'material' at all.[332]

The kind of non-dualistic, 'methodological naturalism' Hinzen proposes for the study of mind if often mistakenly allied with metaphysical doctrines like physicalism.[333] But 'there are no ontological commitments flowing from the way our mind is built, or from how we talk' about events and objects, he adds elsewhere.[334] On the interface between the neural sciences and linguistics, Hinzen again refutes the need for those ontological commitments and questions which Chomsky regards as mere 'harassment':

> The abstract generative categorizations that generative grammar provides are meant to be true of the mind/brain, but do not commit us to describing the reality thus characterized in the very same terms as used in the abstract theory. At the level of cells, very likely, much of our descriptive apparatus will prove inadequate. In the meantime, nothing prevents us from taking our characterizations to be true of structures represented in the mind, rather than representations of these structures.[335]

But we should also be demanding with David Poeppel that neuroscientists provide an explanation for the fundamental entities posited by theoretical linguists and other 'softer' sciences.[336] Science should work both ways, as was the case during the unification of quantum mechanics and molecular theory at the beginning of the twentieth century. If they like, the philosophers can label certain physical theories with metaphysical positions ('realistic materialist monism,' in Galen Strawson's most recent formulation of 'physical reality'), but such questions and games are hardly more than a pest: to entertain the graduate students, and nothing more. We should heed Marr's caution against drawing assumptions about cognitive systems from neurophysiological findings without 'a clear idea about what information needs to be implemented.'[337]

Mark Johnson's recent work stresses the point that 'Traditional logic treats of concepts (i.e., concepts of objects, properties, and relations), propositions, and formal relations. Qualities, if they are mentioned at all, are represented by symbolic placeholders, such as $F(x)$, which is read as 'object x has property (here a quality) F.' Even worse, properties are often regarded as fixed structures 'possessed' by objects, independent of thought.'[338] This stress on normativity in the philosophy of logic was strengthened by the attempt of 'getting clear about the nature of reality through getting clear about the forms of our thoughts or talk about it.'[339] Johnson believes that 'most contemporary logicians usually translate *but* with *and* – they strip away the peculiar felt quality in order to focus only on the relation of connection between x and y. Since modern logic does not and cannot recognize a role for feeling, it must ignore anything but 'pure' formal relation. Consequently, it interprets logical relations as empty formal relations lacking and felt connection or direction in our thinking.'[340]

Though our intuition may concur with Rorty that the mind is a representational device, a 'mirror of nature' and a vessel of truth-conditional contents, its formal structure may disagree.[341] More generally, 'twentieth-century analytic philosophy has been consistently pursuing a path that is in many ways opposed to the broadly rationalist and internalist assumptions about human language and mind' advocated by the biolinguistic enterprise.[342] Like Johnson, Hinzen makes note of an aspect of twentieth century philosophy which has served as an impasse to furthering the mind sciences, and which is worth quoting in full:

> To this day philosophy is centred on man's rationality: his beliefs, desires, and intentions, in terms of which human actions are rationalized. The prime preoccupations of philosophy are the *normatively evaluable* features of the mind: things like when it is correct to assert something, what makes something true, or when a belief is justified. These features are at the same time the main obstacles to viewing the mind as part of the natural world. The puzzle is how to reconcile the intrinsic normativity of mind, meaning, and morality with the world as described by modern science, in which rationality appears to have no place. There is a "realm of law" and a "realm of reason", and they do not fit, giving rise to reductionism, anomalous monism, and other such attempts to close what appears to be a deep divide.[343]

Contrary to Quine's scientism (which afforded a higher priority to the physical sciences than others in describing reality), late twentieth-century analytic philosophy holds that there are many 'languages' attempting to describe the world, including the language of science, though none of them has any stronger claim to the truth than any other: 'This leaves us with a version of philosophy that makes no claims to produce *necessary* truths, be it in the sense of traditional metaphysics or "truths of reason", or in the sense of 'truths of meaning' as opposed to "truths or facts". But it is a philosophy that also does not produce *contingent* empirical truths, on a par with the sciences.' Rejecting a naturalistic 'mentalism,' contemporary analytic philosophy and logic hearkens back to Frege's 'commitment to the objectivity and normativity of thought. ... Given a basic anti-psychologistic commitment that philosophers in the Fregean tradition share, these philosophers have largely focused on *mind-external* entities (thoughts, propositions, referents), while the human mind as empirical domain for naturalistic inquiry has received little attention.'[344]

Arguing instead for a conceptualist view of reference and truth, Jackendoff explained in 1998:

> I am inclined to think that when dealing with the mind, formal parsimony, while not to be neglected, is a second-order consideration. The mind is basically an engineering hack built up piecemeal by evolution, and in my opinion the standard formal tools of set theory and quantificational logic are idealizations that do not do justice to its full richness.[345]

Note that 'formal parsimony' should not be confused with mathematical parsimony (or 'elegance,' to use the term of the physicists), with theoretical 'beauty' being granted the highest importance, following standard scientific practice as recognised by the minimalist (biolinguistic) program.

Carl Becker, the American historian, declared in a lecture given at Stanford University in 1935 that 'The significance of man is that he is that part of the universe that asks the question, What is the significance of Man? He alone can stand apart imaginatively and, regarding himself and the universe in their eternal aspects, pronounce a judgment: The significance of man is that he is insignificant and aware of it.' He continued:

> Man, says Pascal, has this superiority: He knows that the universe can with a breath destroy him, yet at the moment of death he knows that he dies, and knows also the advantage with which the universe thereby has over him; but of all that the universe knows nothing.
> Of all that, the universe knows nothing. Apart from man, the universe knows nothing at all – nothing of itself or of infinite spaces, nothing of man or of his frustrated aspirations, nothing of beginnings or endings, of progress or retrogression, of life or death, of good or evil fortune. The cosmic view of the universe of infinite spaces, and of man's ultimate fate within it, is man's

achievement – the farthest point yet reached in the progressive expansion of human intelligence and power.[346]

Noting the Russellian scopes and limits of naturalistic inquiry, Max Planck also offers a final sobering thought on the limits of the cognitive and natural sciences: 'Science cannot solve the ultimate mystery of nature. And that is because, in the last analysis, we ourselves are part of nature and therefore part of the mystery that we are trying to solve.'[347] In this new internalist environment, it only makes sense to reverse the common adage that consciousness is 'the brain's experience of itself,' and conclude with the anarchist Élisée Reclus that 'Man is nature becoming aware of itself.'[348]

REFERENCES

[1] Cited in Daniela Isac and Charles Reiss, *I-Language: An Introduction to Linguistics as Cognitive Science* (Oxford University Press, 2008), pp. 281-2.
[2] Aldous Huxley, *The Doors of Perception and Heaven and Hell* (London: Flamingo, 1994 [1954]), p. 67.
[3] P. F. Strawson, 'On Referring,' *Mind*, 59(235), July 1950: 332 (emphasis his).
[4] Ibid., p. 339 (emphasis his).
[5] David Michael Levin, *The Philosopher's Gaze: Modernity in the Shadows of the Enlightenment* (University of California Press, 1999), pp. 52-3.
[6] Noam Chomsky, *The Science of Language: Interviews with James McGilvray* (Cambridge University Press, 2012), p. 209.
[7] Paul Pietroski, 'Meaning before Truth,' *Contextualism in Philosophy*, eds. G. Preyer and G. Peter (Oxford University Press, 2002), pp. 253-300.
[8] Noam Chomsky, *Powers and Prospects: Reflections on Human Nature and the Social Order* (London: Pluto, 1996), p. 22.
[9] Ray Jackendoff, *Foundations of Language: Brain, Meaning, Grammar, Evolution* (Oxford University Press, 2002), p. 25.
[10] James McGilvray, 'Meaning and Creativity,' *The Cambridge Companion to Chomsky*, ed. James McGilvray (Cambridge University Press, 2005), p. 213.
[11] Ernest Lepore, 'Donald Davidson,' *A Companion to Analytic Philosophy*, ed. A. P. Martinich and David Sosa (Oxford: Blackwell, 2001), p. 297.
[12] Joshua Hoffman and Gary S. Rosenkrantz, 'Platonic theories of universals,' *The Oxford Handbook of Metaphysics*, ed. Michael J. Loux and Dean W. Zimmerman (Oxford University Press, 2003), p. 46.
[13] James Pustijovsky, *The Generative Lexicon* (Massachusetts: MIT Press, 1995).
[14] Wolfram Hinzen, *Mind Design and Minimal Syntax* (Oxford University Press, 2006), p. 42.
[15] D'Arcy Thompson, *On Growth and Form* (Cambridge University Press, 1917), p. 5.
[16] Lila Gleitman, 'The Learned Component of Language Learning,' *Of Minds and Language: A Dialogue with Noam Chomsky in the Basque Country*, ed. Massimo Piatello-Palmarini, Juan Uriagereka and Pello Salaburu (Oxford University Press, 2009), pp. 240-1 (239-55).

[17] Jackendoff, *Foundations of Language*, p. 377.
[18] Henry James, *Henry James: A Life in Letters*, ed. P. Horne (London: Penguin, 1999/1864-1915), pp. 562-3.
[19] Alex Drummond and Norbert Hornstein, 'Basquing in Minimalism,' review of *Of Minds and Language*, ed. Massimo Piatelli-Palmarini, Juan Uriagereka and Pello Salaburu, *Biolinguistics*, 5(4), 2011: 361 (347-65).
[20] *The Deer Hunter*, dir. Michael Cimino, Universal Pictures, 1978.
[21] Jürgen Trabant, *Humboldt ou le sens du langage* (Liège: Mardaga, 1992).
[22] Pierre Jacob, 'The Scope and Limits of Chomsky's Naturalism,' *Chomsky Notebook*, ed. Jean Bricmont and Julie Franck (Columbia University Press, 2010), pp. 230-1.
[23] Wolfram Hinzen, *An Essay on Names and Truth* (Oxford University Press, 2007), p. 2.
[24] David Lewis, *Convention* (Harvard University Press, 1969), p. 1.
[25] P. F. Strawson, 'Truth,' *Aristotelian Society*, suppl., 24, 1950: 168 (129-56).
[26] George Lakoff, *Women, Fire, and Dangerous Things: What Categories Reveal about the Mind* (University of Chicago Press, 1987), p. 9.
[27] Chomsky, 'The Mysteries of Nature: How Deeply Hidden?' *Chomsky Notebook*, p. 24.
[28] Strawson, 'On Referring,' *Mind*, 340.
[29] Ibid., 326, 327 (emphasis his).
[30] Ibid., 333, 344.
[31] Jackendoff, *Foundations of Language*, p. 35; see Herbert Clark, *Using Language* (Cambridge University Press, 1996).
[32] Jacob, 'Chomsky, Cognitive Science, Naturalism and Internalism,' 2001, http://hal.inria.fr/docs/00/05/32/33/PDF/ijn_00000027_00.pdf, p. 39.
[33] Ibid., pp. 44-5.
[34] Hilary Putnam, 'Replies,' *Philosophical Topics*, 20, 1992: 347-408 (citing his 1975 paper).
[35] Noam Chomsky, *New Horizons in the Study of Language and Mind* (Cambridge University Press, 2000), p. 128.
[36] Mark Johnson, *The Meaning of the Body: Aesthetics of Human Understanding* (University of Chicago Press, 2007), p. 9.
[37] James McGilvray, introduction to Noam Chomsky, *Cartesian Linguistics: A Chapter in the History of Rationalist Thought*, 2nd ed. (Christchurch, New Zealand: Cybereditions, 2002), p. 22.
[38] John Dewey, 'Qualitative Thought,' *The Later Works, 1935-1953*, ed. Jo Ann Boydston (Carbondale: Southern Illinois University Press, 1930/1988), p. 246.
[39] Boban Arsenijević and Wolfram Hinzen, 'Recursion as a Human Universal and as a Primitive,' *Biolinguistics*, 4(2-3), 2010: 166 (165-173).
[40] Robert C. Berwick and Noam Chomsky, 'The Biolinguistic Program: The Current State of its Development,' *The Biolinguistic Enterprise: New Perspectives on the Evolution and Nature of the Human Language Faculty*, ed. Anna Maria di Sciullo and Cedric Boeckx (Oxford University Press, 2011), p. 39.
[41] Wolfram Hinzen, 'Emergence of a Systematic Semantics through Minimal and Underspecified Codes,' *The Biolinguistic Enterprise*, p. 420.
[42] Fernand Braudel, *Civilization and Capitalism 15th-18th Century, Volume I: The Structures of Everyday Life: The Limits of the Possible*, trans. Siân Reynolds (University of California Press, 1992/1979), p. 227
[43] James W. Underhill, *Humboldt, Worldview and Language* (Edinburgh University Press, 2009), p. 41.
[44] Hinzen, *An Essay on Names and Truth*, pp. 83-4.
[45] Ibid., p. 84.
[46] Ibid., p. 84, n. 7.
[47] Cited in Isac and Reiss, *I-Language*, pp. 281-2.
[48] David Marr, *Vision* (San Francisco: Freeman, 1982), p. 270.

[49] Pierre Jacob and Marc Jeannerod, *Ways of Seeing: The Scope and Limits of Visual Cognition* (Oxford University Press, 2003), p. 140.
[50] Akeel Bilgrami and Carol Rovane, 'Mind, Language, and the Limits of Inquiry,' *The Cambridge Companion to Chomsky*, p. 182.
[51] Ibid., p. 184 (emphasis his).
[52] Chomsky, *New Horizons in the Study of Language and Mind*, p. 151.
[53] Hinzen, *Mind Design and Minimal Syntax*, p. ix.
[54] Ibid., pp. ix-x.
[55] Andrew Melnyk, *A Physicalist Manifesto: Thoroughly Modern Materialism* (Cambridge University Press, 2003), pp. 35-6.
[56] Paul M. Pietroski, 'Meaning before Truth,' *Contextualism in Philosophy: Knowledge, Meaning, and Truth*, ed. Gerhard Preyer and Georg Peter (Oxford University Press, 2005), p. 266.
[57] Ibid., p. 284.
[58] Ray Jackendoff, *A User's Guide to Thought and Meaning* (Oxford University Press, 2012), p. 20.
[59] Friedrich Nietzsche, *Writings from the Late Notebooks*, ed. Rüdiger Bittner, trans. Kate Sturge (Cambridge University Press, 2003), p. 14.
[60] Chomsky, *Powers and Prospects*, p. 45.
[61] Ibid., p. 51.
[62] Cited in ibid., p. 51.
[63] Richard Dawkins, *The Selfish Gene*, 2nd ed. (Oxford University Press, 1989).
[64] Francis Crick, *The Astonishing Hypothesis: The Scientific Search for the Soul* (New York: Charles Scribner's Sons, 1994), p. 3.
[65] John Searle, 'Minds and brains without programs,' *Mindwaves*, ed. Colin Blakemore and Susan Greenfield (Oxford: Blackwell, 1987), pp. 209-33.
[66] Jeffrey Gray, *Consciousness: Creeping Up on the Hard Problem* (Oxford University Press, 2004), p. 39.
[67] John Collins, 'Naturalism in the Philosophy of Language; or Why There Is No Such Thing as Language,' *New Waves in Philosophy of Language*, ed. Sarah Sawyer (London: Palgrave Macmillan, 2010), p. 49.
[68] Steven Pinker, *The Stuff of Thought: Language as a Window into Human Nature* (London: Penguin, 2008), p. 24.
[69] Chomsky, *The Science of Language*, pp. 74-5.
[70] Ibid., p. 260.
[71] Friedrich Nietzsche, *Daybreak*, trans. R. Hollingdale (Cambridge University Press, 1997/1881), §115.
[72] Chomsky, *The Science of Lanuage*, p. 260.
[73] Drummon and Hornstein, 'Basquing in Minimalism,' 361.
[74] Cited in Isac and Reiss, *I-Language*, p. 9.
[75] Antoine de Saint Exupéry, *Pilote de Guerre* (1942).
[76] Pinker, *The Stuff of Thought*, p. 287.
[77] Galen Strawson, 'The Self,' *The Oxford Handbook of Philosophy of Mind*, ed. Brian P. McLaughlin, Ansgar Beckermann and Sven Walter (Oxford University Press, 2009), p. 545, 549.
[78] Philip Carr, Review of New Horizons in the Study of Language and Mind by Noam Chomsky, *Journal of Linguistics*, 38(1), March 2002: 149 (147-52).
[79] Arthur Eddington, *The Nature of the Physical World* (Cambridge University Press, 1927).
[80] Hinzen, *Mind Design and Minimal Syntax*, p. 45.
[81] Ibid., p. 47.
[82] Gordon C. Brittan Jr., 'Towards a Theory of Theoretical Objects,' *PSA: Proceedings of the Biennial Meeting of the Philosophy of Science Association*, 1986(1): 384 (384-393).

[83] Peter Ludlow, *The Philosophy of Generative Linguistics* (Oxford University Press, 2011).
[84] Jackendoff, *Foundations of Language*, p. 280.
[85] Bertrand Russell and A. N. Whitehead, *Principia Mathematica* (Cambridge University Press, 1910), p. 66.
[86] Thomas Baldwin, 'Bertrand Russell,' *A Companion to Analytic Philosophy*, p. 33.
[87] John Locke, *An Essay Concerning Human Understanding*, ed. A. D. Woozley (Glasgow: William Collins Sons & Co Ltd, 1964, repr. 1984), p. 259.
[88] John Buridan, *John Buridan: Summulae de dialectica*, trans. Gyula Klima (Yale University Press, 2001), p. 222.
[89] Cited in Bilgrami and Rovane, 'Mind, language, and the limits of inquiry,' *The Cambridge Companion to Chomsky*, p. 189.
[90] Cited in Ramon Selden, Peter Widdowson and Peter Brooker, *A Reader's Guide to Contemporary Literary Theory*, 5th ed. (Harlow: Pearson, 2005), p. 165
[91] Francis Bacon, *Novum Organum*, Book I, *The Works of Francis Bacon Lord Chancellor of England: A New Edition*, ed. and trans. Basil Montague (London, 1831), p. 36.
[92] Nirmalangshu Mukherji, *The Primacy of Grammar* (Massachusetts: MIT Press, 2010).
[93] Cited in Robert C. Berwick and Noam Chomsky, 'The Biolinguistic Program: The Current State of its Development,' *The Biolinguistic Enterprise*, p. 39.
[94] C. R. Gallistel, *The Organization of Learning* (MIT Press, 1990).
[95] C. R. Gallistel, 'Representations in Animal Cognition: An Introduction,' *Animal Cognition*, ed. C. R. Gallistel, *Cognition*, 37, November 1990 (special issue): 1-22.
[96] Anna Maria Di Sciullo, Massimo Piattelli-Palmarini, Kenneth Wexler, Robert C. Berwick, Cedric Boeckx, Lyle Jenkins, Juan Uriagereka, Karin Stromswold, Lisa Lai-Shen Cheng, Heidi Harley, Andrew Wedel, James McGilvray, Elly van Gelderen & Thomas G. Bever, 'The Biological Nature of Human Language,' *Biolinguistics*, 4(1), 2010: 9 (4-34).
[97] W. V. O. Quine, *Ontological Relativity and Other Essays* (New York: Columbia University Press, 1969), p. 26.
[98] Wolfram Hinzen, 'Emergence of a Systematic Semantics through Minimal and Underspecified Codes,' *The Biolinguistic Enterprise*, p. 417.
[99] Ibid., p. 418.
[100] Cited in Hinzen, *Mind Design and Minimal Syntax*, p. 134.
[101] Maggie Tallerman, 'Protolanguage,' *The Oxford Handbook of Language Evolution*, ed. Maggie Tallerman and Kathleen R. Gibson (Oxford University Press, 2012), p. 487 (pp. 479-91).
[102] Chomsky**Error! Bookmark not defined.**, *The Science of Language*, p. 166, n. 3.
[103] Hinzen, *Mind Design and Minimal Syntax*, p. 42.
[104] John W. Yolton, 'Mirrors and Veils, Thoughts and Things: The Epistemological Problematic,' *Reading Rorty*, ed. A. R. Malachowski (Oxford: Blackwell, 1990), p. 62 (pp. 58-73).
[105] Gilbert Harman, Review of New Horizons in the Study of Language and Mind by Noam Chomsky, *The Journal of Philosophy* 98(5), May 2001: 265 (265-9).
[106] Robert C. Berwick and Noam Chomsky, 'The Biolinguistic Program: The Current State of its Development,' *The Biolinguistic Enterprise*, p. 40.
[107] Wolfram Hinzen, 'Emergence of a Systematic Semantics through Minimal and Underspecified Codes,' *The Biolinguistic Program*, p. 419.
[108] Ernie Lepore and Kirk Ludwig, *Donald Davidson: Meaning, Truth, Language, and Reality* (Oxford University Press, 2005), p. 79.
[109] Noam Chomsky, Review of *Meaning and Use* by Rulon Wells, *The Journal of Symbolic Logic*, 22(1), March 1957: 87-8.

[110] David Hume, *An Enquiry Concerning Human Understanding*, in *Enquiries Concerning the Human Understanding and Concerning the Principles of Morals*, ed. L. A. Selby-Bigge, 2nd ed. (Oxford: Clarendon Press, 1902), Section XII, 'Of the academical or sceptical Philosophy,' Part III, p. 165.

[111] Galen Strawson, *Real Materialism and Other Essays* (Oxford University Press, 2008), p. 422.

[112] Cory Juhl and Eric Loomis, *Analyticity* (London: Routledge, 2010), p. 270.

[113] Charles Barber, *The English Language: A Historical Introduction* (Cambridge University Press, 1993), p. 2.

[114] Donald Davidson, 'The Emergence of Thought,' *Subjective, Intersubjective, Objective*, ed. Donald Davidson (Oxford University Press, 2001), p. 123.

[115] Jerry Fodor, *Concepts: Where Cognitive Science Went Wrong*, Oxford Cognitive Science Series (Oxford: Clarendon Press, 1998), p. 28.

[116] Bill Brewer, 'Perception and Its Objects,' *Philosophical Studies: An International Journal for Philosophy in the Analytic Tradition*, 132(1), January 2007, Selected Papers from the American Philosophical Association, Pacific Division, 2006 Meeting: 87 (87-97).

[117] Hilary Putnam, 'Sense, Nonsense, and the Senses: An Inquiry into the Powers of the Human Mind,' *The Journal of Philosophy*, 91, 1994: 445-517.

[118] John McDowell, 'Criteria, Defeasibility, and Knowledge,' 1982, *Perceptual Knowledge*, ed. J. Dancy (Oxford University Press, 1988), pp. 409-19.

[119] David Lewis, 'Languages and Language,' *Language, Mind, and Knowledge*, ed. Keith Gunderson (University of Minnesota Press, 1975), pp. 163-88.

[120] Pietroski, 'Meaning and Truth,' *Contextualism in Philosophy*.

[121] Strawson, *Real Materialism and Other Essays*, p. 65.

[122] Alex Byrne and James Pryor, 'Bad Intentions,' *Two-Dimensional Semantics*, eds. Manuel García-Carpintero and Josep Macià (Oxford University Press, 2006), p. 38, n. 1.

[123] Ibid., p. 41.

[124] François Recanti, 'Indexical Concepts and Compositionality,' *Two-Dimensional Semantics*, p. 250.

[125] Gillian Russell, 'Analyticity in Externalist Languages,' *New Waves in Philosophy of Language*, p. 190 (186-205).

[126] James R. Hurford, 'The Origins of Meaning,' *The Oxford Handbook of Language Evolution*, ed. Maggie Tallerman and Kathleen R. Gibson (Oxford University Press, 2012), p. 371 (pp. 370-81).

[127] Jacob and Jeannerod, *Ways of Seeing*, p. 3.

[128] Ibid., p. 4.

[129] Ibid., p. 55. See Mortimer Mishkin, 'Cortical visual areas and their interactions,' *Brain and Human Behavior*, ed. A. C. Karczmar and J. C. Eccles (Berlin: Springer, 1972), pp. 187-208.

[130] David Hume, *A Treatise of Human Nature*, ed. L. A. Selby-Bigge and P. H. Nidditch (Oxford: Clarendon Press, 1978/1739-40), p. 68 (emphasis his).

[131] Hume, *A Treatise of Human Nature*, p. 241.

[132] Hume, *An Enquiry Concerning Human Understanding*, p. 33 (emphasis his).

[133] Ibid., p. 30 (emphasis his).

[134] Richard Rorty, *Consequences of Pragmatism: Essays 1972-1980* (University of Minnesota Press, 1982), p. xxiii.

[135] Chomsky, *New Horizons in the Study of Language and Mind*, p. 132, quoting Noam Chomsky, *Syntactic Structures* (The Hague: Mouton, 1957), pp. 102-3.

[136] Steven Pinker, *The Language Instinct: How the Mind Creates Language* (London: Harper Collins, 1994), p. 126.

[137] Friedrich Albert Lange, *The History of Materialism and Criticism of its Present Importance*, trans. Ernest Chester Thomas, 3rd ed. (London: Routledge & Kegan Paul, 1957), pp. 53-4.
[138] Ibid., p. 322.
[139] Hinzen, *Mind Design and Minimal Syntax*, p. x.
[140] Jeff Speaks, 'Introduction, Transmission, and the Foundations of Meaning,' *New Waves in Philosophy of Language*, p. 244 (226-49).
[141] Hinzen, *Mind Design and Minimal Syntax*, p. 138.
[142] Hinzen, *An Essay on Names and Truth*, p. 108.
[143] Karola Stotz and Paul Griffiths, 'Biohumanities: Rethinking the Relationship Between Biosciences, Philosophy and History of Science, and Society,' *The Quarterly Review of Biology*, 83(1), 2008: 41 (37-45).
[144] Ram Neta, 'Contextualism and the Problem of the External World,' *Philosophy and Phenomenological Research*, 66(1), January 2003: 1-31.
[145] John Ziman, *Public Knowledge* (Cambridge University Press, 1968), p. 9.
[146] Cited in Hannah Arendt, *The Life of the Mind* (London: Harcourt, 1977), p. 167.
[147] Johnson, *The Meaning of the Body*, p. 47.
[148] Hinzen, *Mind Design and Minimal Syntax*, pp. 66-7.
[149] Bertrand Russell, 'On Denoting,' *Mind*: 14, 1905: 484-5 (479-93).
[150] Carl Hempel, 'Studies in the Logic of Confirmation,' *Mind*, 54(213), January 1945: 1-26, 97-121.
[151] Pinker, *The Stuff of Thought*, p. 292.
[152] Francis Crick, *The Astonishing Hypothesis* (London: Macmillan, 1994), p. 33.
[153] Tom Bever, 'The Cognitive Basis for Linguistic Structures,' *Cognition and the Development of Language*, ed. J. R. Hayes (New York: J. Wiley and Sons, 1970), pp. 279-362.
[154] Donald Davidson, *Inquiries into Truth and Interpretation*, 2nd ed. (Oxford University Press, 2001), p. xv.
[155] Ray Jackendoff, *Language, Consciousness, Culture: Essays on Mental Structure* (Massachusetts: MIT Press, 2007).
[156] Hinzen, *Mind Design and Minimal Syntax*, p. 154, n. 4.
[157] Ibid., pp. 154-5.
[158] Cited in Cedric Boeckx, *Linguistic Minimalism: Origins, Concepts, Methods, and Aims* (Oxford University Press, 2006), p. 4.
[159] Barbara Abbott, 'Models, Truth, and Semantics,' *Linguistics and Philosophy*, vol. 20 (1997): 117-38. See also David Lewis, 'Psychophysical and Theoretical Identifications,' *Australasian Journal of Philosophy*, vol. 50 (December 1972): 249-58; Robert Stalnaker, *Context and Content: Essays on Intentionality in Speech and Thought* (Oxford University Press, 1999); Richard Montague, 'The Proper Treatment of Quantification in Ordinary English,' *Approaches to Natural Language*, eds. Jaakko Hintikka, Julius Moravcsik, Patrick Suppes (Dordrecht: Reidal, 1973) pp. 221-242; B. H. Partee, 'Montague grammar and transformational grammar,' *Linguistic Inquiry*, vol. 6 (1975): 203-300.
[160] Jackendoff, *Foundations of Language*, p. 296.
[161] Locke, *An Essay Concerning Human Understanding*, p. 58.
[162] Derek Gjersten, *Science and Philosophy: Past and Present* (Oxford University Press, 1989), p. 13.
[163] Heinrich Hertz, *Principles of Mechanics*, trans. D. E. Jones and J. T. Walley (London: Macmillan, 1899), pp. 7-8.
[164] James Joyce, *Ulysses*, ed. Jeri Johnson (Oxford University Press, 2008), p. 31.
[165] Aldous Huxley, *The Doors of Perception and Heaven and Hell*, p. 50.
[166] Thomas Nagel, 'What is it like to be a bat?' *The Philosophical Review*, LXXXIII(4), October 1974: 435-50.
[167] Colin McGinn, 'All machine and no ghost?', *New Statesman*, 20 February 2012.

[168] Robert C. Berwick and Noam Chomsky, 'The Biolinguistic Program: The Current State of its Development,' *The Biolinguistic Enterprise*, p. 39.
[169] Hinzen, *Mind Design and Minimal Syntax*, p. 26.
[170] Jacob, 'Chomsky, Cognitive Science, Naturalism and Internalism,' 2001, http://hal.inria.fr/docs/00/05/32/33/PDF/ijn_00000027_00.pdf, p. 46.
[171] Noam Chomsky, 'Explaining language use,' *Philosophical Topics*, vol. 20, pp. 208-9.
[172] Chomsky, *New Horizons in the Study of Language and Mind*, p. 125.
[173] Hinzen, *Mind Design and Minimal Syntax*, p. 259.
[174] McGilvray, 'Meaning and Creativity,' *The Cambridge Companion to Chomsky*, p. 214.
[175] See Noam Chomsky's lecture "On Referring' Revisited' delivered at Harvard University on 30 October 2007, http://a9.video2.blip.tv/9400000011619/Internalism-NoamChomskyOnReferringRevisited552.mov?bri=61.3&brs=709.
[176] Chomsky, *Powers and Prospects*, p. 21.
[177] Pinker, *The Stuff of Thought*, p. 179-80.
[178] Noam Chomsky, 'Language and the Cognitive Science Revolution(s),' lecture given at Carleton University, 8 April 2011, http://chomsky.info/talks/20110408.htm.
[179] Ralph Cudworth, *A Treatise Concerning Eternal and Immutable Morality*, ed. S. Hutton (Cambridge University Press, 1996), p. 91.
[180] Boban Arsenijević, 'From spatial cognition to language,' *Biolinguistics*, 2(1), 2008: 15 (3-23).
[181] Jackendoff, *Foundations of Language*, p. 323.
[182] Hinzen, *An Essay on Names and Truth*, p. 79.
[183] Ibid., p. 81.
[184] Rodolfo Llinás, "Mindness' as a functional state of the brain,' *Mindwaves: Thoughts on Intelligence, Identity and Consciousness*, ed. Colin Blakemore and Susan Greenfield (Oxford: Blackwell, 1987), p. 344.
[185] Ibid., p. 351.
[186] Hermann von Helmholtz, *Handbuch der Physiologischen Opik*, Zweite Auflag (Hamburg: Voss, 1896).
[187] Llinás, "Mindness' as a functional state of the brain,' *Mindwaves*, pp. 353-4.
[188] James Higginbotham, 'Two Interfaces,' 'Discussion,' *Of Minds and Language*, p. 152 (142-154).
[189] Stephen Barker, *Global Expressivism: Language Agency without Semantics, Reality without Metaphysics* (University of Nottingham ePrints, 2007), p. 302, available at http://eprints.nottingham.ac.uk/696/.
[190] Peter Ludlow, 'Noam Chomsky,' *A Companion to Analytic Philosophy*, p. 425.
[191] Noam Chomsky, *Lectures on Government and Binding* (Dordrecht: Foris Publications, 1981), p. 324.
[192] McGilvray, 'Meaning and Creativity,' *The Cambridge Companion to Chomsky*, p. 214.
[193] Ludwig Wittgenstein, *Philosophical Investigations*, §129.
[194] Jacob, 'Chomsky, Cognitive Science, Naturalism and Internalism,' 2001, http://hal.inria.fr/docs/00/05/32/33/PDF/ijn_00000027_00.pdf, p. 43.
[195] Ibid., p. 42.
[196] Strawson, 'The Self,' *The Oxford Handbook of Philosophy of Mind*, p. 543.
[197] Paul Elbourne, *Meaning: A Slim Guide to Semantics* (Oxford University Press, 2011), p. 26.
[198] Ray Jackendoff, 'Parts and boundaries,' *Cognition*, 41, 1991: 9-45.
[199] Stephen Handel, *Listening: An Introduction to the Perception of Auditory Events* (Massachusetts: MIT Press, 1989), cited in Isac and Reiss, *I-Language*, p. 29.
[200] Ray Jackendoff, *Semantics and Cognition* (MIT Press, 1983).
[201] Jackendoff, *Foundations of Language*, p. 339.

[202] Chomsky, *New Horizons in the Study of Language and Mind*, p. 125.
[203] McGilvray, 'Meaning and Creativity,' *The Cambridge Companion to Chomsky*, p. 218.
[204] Hinzen, *An Essay on Names and Truth*, p. 9.
[205] T. Langendoen, P. M. Postal, *The Vastness of Natural Language* (University of Chicago Press, 1984).
[206] Cited in Chomsky, *Cartesian Linguistics*, p. 61.
[207] Jackendoff, *Foundations of Language*.
[208] D. Raffman, 'On the persistence of phenomenology,' *Conscious Experience*, ed. T. Metzinger (Thorverton: Inprint Academic, 1995), p. 295.
[209] Noam Chomsky, *Reflections on Language* (New York: Pantheon, 1975), p. 204.
[210] Pinker, *The Stuff of Thought*, p. 97.
[211] Jackendoff, *Foundations of Language*, p. 275.
[212] Hinzen, *Mind Design and Minimal Syntax*, p. 170, n. 11.
[213] Stephen J. Gould an Elisabeth. S. Vrba, 'Exaptation – A missing term in the science of form,' 1982, repr. in *The Philosophy of Biology*, eds. D. Hull and M. Ruse (Oxford University Press, 1998), p. 55 (pp. 52-71).
[214] Hinzen, *Mind Design and Minimal Syntax*, p. 170, n. 1. See A. Carstairs-McCarthy, *The Origins of Complex Language: An Inquiry into the Evolutionary Beginnings of Sentences, Syllables, and Truth* (Oxford University Press, 1999).
[215] Ronald Langacker, 'Conceptualization, Symbolization, and Grammar,' in Michael Tomasello (ed.), *The New Psychology of Language* (Hillside, New Jersey: Erlbaum, 1998), p. 19.
[216] Jackendoff, *Foundations of Language*, p. 124.
[217] Henry Maudsley, *The Physiology and Pathology of the Mind* (New York: Appleton, 1867), cited in Iain McGilchrist, *The Master and His Emissary: The Divided Brain and the Making of the Western World* (Yale University Press, 2010), p. 53.
[218] Cited in Elbourne, *Meaning*, p. 22.
[219] Rudolf Carnap, *Meaning and Necessity: A Study in Semantics and Modal Logic*, enlarged edition (University of Chicago Press, 1956), p. 11.
[220] David Hume, *A Treatise of Human Nature*, ed. L. A. Selby-Bigge, 2nd ed. (Oxford: Clarendon, 1739-40/1978), I, p. 257.
[221] Chomsky, *New Horizons in the Study of Language and Mind*, p. 150.
[222] Hinzen, *Mind Design and Minimal Syntax*, p. 267.
[223] Ibid., pp. 266-7.
[224] Ibid., p. 276.
[225] Ibid.
[226] Lakoff, *Women, Fire, and Dangerous Things*, p. xii.
[227] Leonard Talmy, 'The Relation of Grammar to Cognition: A Synopsis,' *Theoretical Issues in Natural Language Processing 2*, ed. D. Waltz (New York: Association for Computing Machinery, 1978), pp. 14-24.
[228] Jackendoff, *Foundations of Language*, p. 352.
[229] Cited in Lange, *The History of Materialism and Criticism of its Present Importance*, pp. 41-2.
[230] Chomsky, *New Horizons in the Study of Language and Mind*, p. 16.
[231] Lange**Error! Bookmark not defined.**, *The History of Materialism and Criticism of its Present Importance*, pp. 83-4 (emphasis mine).
[232] Ibid., p. 193.
[233] Ibid., pp. 278-9.
[234] Chomsky, *Powers and Prospects*, p. 22.
[235] MacNamara, *Names for Things* (MIT Press, 1982).
[236] Donald Davidson, 'The Social Aspect of Language,' in B. McGuinness and G. Oliveri (eds.), *The Philosophy of Michael Dummett* (Dordrecht: Kluwer, 1994), p. 11.

[237] Jackendoff, *Foundations of Language*, p. 304.
[238] Ibid., p. 308.
[239] Ibid., p. 307.
[240] *Futurama*, S07E11, 'Viva Mars Vegas,' dir. Frank Marino, 22 August 2012.
[241] Charles J. Fillmore, 'Frame semantics,' *Linguistics in the Morning Calm* (Seoul: Hanshin Publishing Co., 1982), pp. 111-137.
[242] Jackendoff, *Foundations of Language*, p. 354.
[243] Simon Garrod, Gillian Ferrier and Siobhan Campbell, '*In* and *On*: Investigating the Functional Geometry of Spatial Prepositions,' *Cognition*, vol. 72 (1999): 167-89.
[244] Jackendoff, *Foundations of Language*, p. 354.
[245] Ibid.
[246] Ray Jackendoff, *Semantics and Cognition* (Massachusetts: MIT Press, 1982).
[247] Pinker, *The Stuff of Thought*, p. 187.
[248] Jackendoff, *Foundations of Language*, p. 355.
[249] Ibid., p. 371.
[250] Ruth Millikan, *Language and Other Abstract Objects* (Massachusetts: MIT Press, 1984), p. 17.
[251] Jackendoff, *Foundations of Language*, p. 372.
[252] Edward Osborne Wilson, *On Human Nature* (Harmondsworth: Penguin, 1978), p. 195.
[253] Deborah Keleman, 'Are children 'intuitive theists'?', *Psychological Science*, 15: 5, 2004, pp. 295-301.
[254] Richard Dawkins, *The God Delusion* (London: Bantam Press, Black Swan edition, 2007), p. 210.
[255] Marx X. Wartofsky, *Conceptual Foundations of Scientific Thought: An Introduction to the Philosophy of Science* (New York: Macmillan, 1968), p. 48.
[256] Jackendoff, *Foundations of Language*, p. 372.
[257] Chomsky, *Powers and Prospects*, p. 38.
[258] Jackendoff, *Foundations of Language*, p. 328.
[259] Wolfram Hinzen, 'Emergence of a Systematic Semantics through Minimal and Underspecified Codes,' *The Biolinguistic Program*, p. 421
[260] Ibid., p. 422
[261] Wolfram Hinzen, 'Emergence of a Systematic Semantics through Minimal and Underspecified Codes,' *The Biolinguistic Program*, p. 421, n. 1.
[262] Carnap, 'Meaning and Synonymy in Natural Languages,' *Meaning and Necessity*, pp. 238.
[263] Hinzen, *Mind Design and Minimal Syntax*, p. 159.
[264] Jackendoff, *Language, Consciousness, Culture*, pp. 246-7.
[265] Hinzen, *Mind Design and Minimal Syntax*, p. 22.
[266] Elbourne, *Meaning*, p. 1.
[267] McGilchrist, *The Master and His Emissary*, p. 285-6. See Bruno Snell, *The Discovery of the Mind* (New York: Harper & Row, 1960), p. 229.
[268] Wallace Chafe, 'Language and Consciousness,' *The Cambridge Handbook of Consciousness*, ed. Philip David Zelazo, Marris Moscovitch and Evan Thompson (Cambridge University Press, 2007), p. 371.
[269] For an illuminating analysis of adjunction structures see Norbert Hornstein and Jairo Nunes, 'Adjunction, Labeling, and Bare Phrase Structure,' *Biolinguistics*, 2(1), 2008: 57-86.
[270] Hinzen, *Mind Design and Minimal Syntax*, pp. 176-7.
[271] Ibid., p. 178.
[272] Jackendoff, *Foundations of Language*, p. 351.
[273] Hinzen, *Mind Design and Minimal Syntax*, p. 180.
[274] Michael Dummett, *Frege: Philosophy of Language* (London: Duckworth, 1973).

[275] William James, *Pragmatism, A New Name for Some Old Ways of Thinking* (1907/1978), p. 97.
[276] Chomsky, *Powers and Prospects*, p. 52.
[277] Hinzen, *Mind Design and Minimal Syntax*, p. 164.
[278] G. E. Moore, 'Facts and Propositions,' *Aristotelian Society Supplementary Vol. III* (1927), Symposium with Frank P. Ramsey, reproduced in G. E. Moore, *Philosophical Papers* (London: George, Allen and Unwin, 1959), pp. 60-88 (specifically p. 71).
[279] Donald Davidson, 'Truth and Meaning,' *Inquiries into Truth and Interpretation*, 2nd ed. (Oxford University Press, 2001), p. 33, originally published in *Synthese*, 17, 1967: 304-23.
[280] James Higginbotham, *Tense, Aspect, and Indexicality* (Oxford University Press, 2009), p. 11.
[281] Ibid., p. 17.
[282] Hans Reichenbach, *Elements of Symbolic Logic* (London: Macmillan, 1947).
[283] Hinzen, *An Essay on Names and Truth*, p. 26.
[284] Daniel Dennett, *Darwin's Dangerous Idea: Evolution and the Meaning of Life* (New York: Simon & Schuster, 1995).
[285] Zeno Vendler, *Linguistics in Philosophy* (Cornell University Press, 1967).
[286] Hinzen, *An Essay on Names and Truth*, p. 125.
[287] Hinzen, *Mind Design and Minimal Syntax*, p. 190.
[288] Paul Pietroski, 'Function and concatenation,' *Logical Form and Language*, ed. G. Preyer and G. Peter (Oxford University Press, 2002), p. 106 (pp. 91-117).
[289] Hinzen, *Mind Design and Minimal Syntax*, p. 191.
[290] Tanmoy Bhattacharya, Review of *Mind Design and Minimal Syntax* by Wolfram Hinzen, http://www.fosssil.in/fn_web_1/book%20review_hinzen.htm.
[291] Hinzen, *Mind Design and Minimal Syntax*, p. 192.
[292] Ibid., p. 209.
[293] John Bolender, Burak Erdeniz and Cemil Kerimoğlu, 'Human Uniqueness, Cognition by Description, and Procedural Memory,' *Biolinguistics*, 2(2-3), 2008: 132 (129-151)..
[294] Hinzen, *An Essay on Names and Truth*, pp. 41-2.
[295] Ibid., p. 83.
[296] See Jackendoff, *Foundations of Language*, p. 137 for further discussion and examples.
[297] Ibid., p. 181.
[298] Ibid.
[299] Ibid., p. 290; Beth Levin and Malka Rappaport Hovav, 'Lexical Semantics and Syntactic Structure,' *The Handbook of Contemporary Semantic Theory*, ed. Shalom Lappin (Oxford: Blackwell, 1996).
[300] Ibid., p. 268.
[301] Jackendoff, *Foundations of Language*, p. 121.
[302] Steven Pinker, *How the Mind Works* (New York: W. W. Norton & Company, 1997), p. 561.
[303] Noam Chomsky, 'Approaching UG from below,' *Interfaces + Recursion = Language?* eds. Uli Sauerland and Hans Martin Gärtner (New York: Mouton de Gruyter, 2007), p. 15 (1-29).
[304] D. Braun and J. Saul, 'Simple sentences, substitution, and mistaken evaluations,' *Philosophical Studies*, 111(1), 2002: 1-41.
[305] Hinzen, *An Essay on Names and Truth*, pp. 218-9, n. 14.
[306] Ibid., p. 188.
[307] Ibid., p. 198.
[308] Pinker, *The Stuff of Thought*, p. 171; W. V. O. Quine, *Word and Object* (MIT Press, 1960), sect. 19; N. Soja, S. Carey and E. Spelke, 'Ontological categories guide young children's inductions of word meaning: Object terms and subject terms,' *Cognitions*, 38, 1991: 179-21.

[309] Ibid., p. 168.
[310] Ibid., see Richard Lederer, *Crazy English* (New York: Pocket Book, 1990).
[311] Ibid., p. 169, 170.
[312] Ibid., p. 171.
[313] Hinzen, *An Essay on Names and Truth*, p. 90.
[314] Paul Pietroski, Jeffrey Lidz, Tim Hunter and Justin Halberda, 'The Meaning of "Most": Semantics, Numerosity, and Psychology,' *Mind and Language*, 24, 2009: 554-85.
[315] Pinker, *The Stuff of Thought*, p. 199.
[316] Arsenijević, 'From spatial cognition to language,' *Biolinguistics*, 13.
[317] Ibid., 15; Linda Hermer-Vazquez, Elisabeth S. Spelke & Alla Katsnelson, 'Source of flexibility in human cognition: Dual task studies of space and language,' *Cognitive Psychology* 39, 1999: 3-36.
[318] Frank Ramsey, *Philosophical Papers*, ed. D. H. Mellor (Atlantic Highlands, New Jersey: Humanities Press, 1990), pp. 38-9.
[319] Hinzen, *An Essay on Names and Truth*, p. 5.
[320] Thomas Kuhn, *The Structure of Scientific Revolutions* (University of Chicago Press, 1970), p. 206.
[321] John Collins, 'Naturalism in the Philosophy of Language; or Why There Is No Such Thing as Language,' *New Waves in Philosophy of Language*, ed. Sarah Sawyer (London: Palgrave Macmillan, 2010), p. 47.
[322] Ibid., p. 48.
[323] Chomsky, *New Horizons in the Study of Language and Mind*, p. 95.
[324] Mark Baker, *The Atoms of Language* (New York: Basic Books, 2001).
[325] George Wells Beadle and Muriel Beadle, *The Language of Life: An Introduction to the Science of Genetics* (Garden City: Doubleday, 1966), p. 68.
[326] Ernst Mach, *The Science of Mechanics: A Critical and Historical Account of Its Development* (La Salle, Illinois: Open Court, 1962), p. 492.
[327] William James, 'Pragmatism's Conception of Truth,' *Pragmatism: A New Name for Some Old Ways of Thinking* (New York: Longman, 1907), p. 83.
[328] Hinzen, *Mind Design and Minimal Syntax*, p. 5.
[329] Ibid., p. 22.
[330] Cited in Neil Smith, *Chomsky: Ideas and Ideals* (Cambridge University Press, 1999), p. 98.
[331] Cited in Hinzen, *Mind Design and Minimal Syntax*, p. 7.
[332] Ibid., p. 8.
[333] J. Kim, *Physicalism or Something Near Enough* (Princeton University Press, 2005), D. Chalmers and F. Jackson, 'Conceptual Analysis and Reductive Explanation,' *Philosophical Review*, 110: 316, 60.
[334] Higginbotham, 'Two Interfaces,' 'Discussion,' *Of Minds and Language*, p. 153.
[335] Hinzen, *Mind Design and Minimal Syntax*, p. 84.
[336] David Poeppel, Interview with Roger Bingham at CogSci 2011, 23 July 2011: http://thesciencenetwork.org/programs/cogsci-2011/interview-with-david-poeppel.
[337] David Marr, *Vision: A Computational Investigation into the Human Representation of Processing of Visual Information* (W. H. Freeman, 1982), p. 26.
[338] Johnson, *The Meaning of the Body*, p. 72.
[339] Ernest Lepore and Kirk Ludwig, 'What is logical form?', *Logical Form and Language*, eds. Gerhard Preyer and Georg Peter (Oxford University Press, 2002), p. 55 (pp. 54-90).
[340] Johnson, *The Meaning of the Body*, p. 96.
[341] Richard Rorty, *Philosophy and the Mirror of Nature* (Princeton University Press, 1980).
[342] Hinzen, *Mind Design and Minimal Syntax*, p. 32.

[343] Ibid., p. 33.
[344] Ibid., p. 34, 40.
[345] Ray Jackendoff, 'Why a Conceptualist View of Reference? A Reply to Abbott,' *Linguistics and Philosophy*, 21(2), April 1998: 217-8 (211-9).
[346] Carl Lotus Becker, *Progress and Power* (Stanford University Press, 1935), pp. 100-102.
[347] Max Planck, *Where is Science Going?*, trans. James Murphy (London: Allen & Unwin, 1933), p. 217.
[348] McGilchrist, *The Master and His Emissary*, p. 19; Élisée Reclus, 'L'homme est la nature prenant conscience d'elle-même,' in *L'homme & la terra* (1908), ed. FM/La Découverte (1982), p. 1.

INDEX

A Bit of Fry & Laurie, 25
Abbey, Edward, 84
Abbott, Barbara, 216
Abelard, 46
Agassiz, Louis, 49
Allen, Richard C., 142
Allen, Woody, 178
Anderson, Philip, 129
Annie Hall, 101
Anshen, Ruth Nanda, 20
Antoine de Saint Exupéry, 204
Antoine de Saint-Exupéry, 45
Antony, Louise M., 23
Appleton, Edward, 177
Aristotle, 27, 70, 83, 97, 147, 193, 194, 201, 202, 217, 228, 230, 244
Arndt, Markus, 115
Arnheim, Rudolf, 158
Arsenijević, Boban, 245
Arshavsky, V. V., 29
Atkins, Peter, 148
Aubrey, John, 136
Austin, John, 217, 235
Ayers, Michael, 165
Bacon, Francis, 46, 207
Baer, Karl Ernst von, 49
Baldwin, Thomas, 206
Barber, Charles, 210
Barker, Stephen, 222
Barrow, John D., 134
Beadle, George, 246
Beadle, Muriel, 246
Becker, Carl, 250
Berlin, Isaiah, 30
Bernard, Claude, 36
Berwick, Robert, 24, 32, 33, 35, 36, 57, 58, 63, 64, 198, 209, 218
Bever, Tom, 215

Bilgrami, Akeel, 200
Blake, William, 47
Bloom, Paul, 16, 48
Boas, Franz, 115
Boeckx, Cedric, 28, 49, 52, 57
Bohm, David, 160
Bohr, Niels, 82
Bolt, Usain, 149
Boltzmann, Ludwig, 140
Born, Max, 128
Boroditsky, Lera, 115
Bošković, Ruđer, 4
Bower, Thomas, 96
Brenner, Sydney, 86
Brewer, Bill, 210
Broca's area, 120
Broca's area, 92, 120
Brooks, David, 176
Buridan, John, 206
Cacioppo, John, 103
Campbell, Joseph, 136
Campbell, Siobhan, 233
Carnap, Rudolf, 120, 133, 236
Carr, Philip, 205
Carroll, Lewis, 140
Carroll, Sean, 124
Carruthers, Peter, 96
Carstairs-McCarthy, Andrew, 227
Chalmers, David, 79, 211
Chaminade, Thierry, 174
Chaucer, Geoffrey, 34
Chekhov, Anton, 4
Chesterton, G. K., 163, 177
Chierchia, Gennaro, 17
Chomsky, Noam, 6, 8, 9, 10, 11, 12, 14, 15, 16, 20, 21, 23, 24, 26, 28, 31, 32, 33, 34, 35, 36, 49, 52, 56, 57, 58, 61, 62, 63, 66, 67, 68, 70, 78, 79, 81, 85, 87, 91, 97, 101, 103, 104, 111,

120, 125, 128, 129, 131, 132,
138, 140, 142, 146, 152, 156,
157, 193, 195, 198, 200, 205,
206, 208, 209, 210, 213, 216,
218, 219, 220, 222, 223, 225,
226, 227, 228, 229, 230, 231,
232, 235, 238, 243, 244, 246,
247, 248
Chopin, Frédéric, 177
Chopra, Deepak, 84
Church, Alonzo, 23
Churchland, Patricia, 129
Clark, Herbert, 197
Coetzee, J. M., 37
Coleridge, Samuel Taylor, 81, 138
Collins, John, 121, 246
Colombia, 150
Comte, Auguste, 136
Conrad, Joseph, 54
Copernicus, Nicolaus, 46, 130
Corballis, Michael, 49
Cosmides, Leda, 98
Craik, Kenneth, 169
Crick, Francis, 203, 215
Crysis 2, 164
Cudworth, Ralph, 220
Daly, Chris, 129
Damasio, Antonio, 19, 170
Darwin, Charles, 6, 49, 50, 53, 65,
78, 132, 140, 156, 215
Davidson, Donald, 24, 31, 104,
163, 193, 210, 215, 231
Davie, George, 6
Davies, Martin, 34
Davy, Humphry, 135
Dawkins, Richard, 48, 51, 66, 68,
134, 203
De Broglie, Louis, 128
De Niro, Robert, 194
Decety, Jean, 174
Deglin, V. L., 161
Delacroix, Eugène, 158
Democritus, 147, 197
Dennett, Daniel, 11, 65, 66, 68,
78, 79, 85, 102, 134, 150, 160,
163, 171, 239, 247
Derrida, Jacques, 84

Descartes, Rene, 4, 13, 31, 90, 93,
101, 112, 125, 126, 132, 137,
138, 149, 162, 163, 170, 175,
191, 208, 222, 225
Deutsch, David, 171
Dewey, John, 31, 140, 162, 164
Diderot, Denis, 78
Dirac, Paul, 45, 133, 142
Dittrich, Winand, 100
Dobzhansky, Theodosius, 51, 53,
78, 172
Douglas, Kate, 78
Dretske, Fred, 211
Drummond, Alex, 204
Dummett, Michael, 11, 238
Duncan, John, 92, 116, 149, 150,
153
Dunn, Elizabeth W., 154
Dupré, John, 86
Dyson, Freeman, 150
Eddington, Arthur, 7, 136
Edelman, Gerald, 153
Einstein, Albert, 23, 118, 121,
123, 128, 142, 150, 151
Elbourne, Paul, 223, 237
Elizabeth Anscombe, 6
Elwes, Richard, 118
Epictetus, 13
Epicurus, 131
Epley, Nicholas, 103
Feinberg, Gerald, 127, 147, 162
Ferrier, Gillian, 233
Fillmore, Charles, 233
Fisher, Ronald, 51
Fitch, William Tecumseh, 20, 26
Flew, Anthony, 131
Fodor, Jerry, 210, 227
Foss, Martin, 169
Foucault, Michel, 66, 67
Fox-Keller, Evelyn, 47
Fraassen, Ban van, 78
Frege, Gottlob, 22, 23, 198, 202,
209, 212, 218, 244, 250
Freud, Sigmund, 176
Friedman, Michael, 144
Fromkin, Victoria, 18
Fry, Stephen, 25
Futurama,, 232

Galilei, Galileo, 4, 52, 53, 61, 83, 93, 131, 132, 147, 247
Gallistel, Randy, 113, 153
Garrett, Don, 18
Garrod, Simon, 233
Gatsby, Jay, 153, 228
Gaukroger, Stephen, 126
Geary, David, 80
Geddes, Linda, 124
Gefter, Amanda, 151
Gell-Mann, Murray, 120, 138
Gelman, Rochel, 57
Gentner, Dedre, 118
Gibbs, Raymond, 113
Gide, André, 5
Gjersten, Derek, 133, 135, 217
Gleitman, Lila, 100, 194
Goatly, Andrew, 111
Gödel, Kurt, 145, 241
Goethe, 90
Goethe, Johann Wolfgang von, 49, 89
Goldschmidt, Richard, 49
Goodall, Jane, 207
Goodman, Nelson, 202
Goodwin, Brian, 49
Gopnik, Alison, 99
Gottlieb, Anthony, 87
Graham-Rowe, Duncan, 92
Gray, Jeffrey, 85, 87
Greece, 237
Greene, Brian, 124
Griffiths, Paul, 214
Griggs, Jessica, 158
Guthrie, Stewart, 102
Guttenplan, Samuel, 156, 171
Hague, William, 125, 236
Haldane, J. B. S., 146
Haman, Johan, 32
Hameroff, Stuart, 85
Handel, Stephen, 224
Harris, Sam, 78
Hassabis, Dennis, 26
Hauser, Marc, 20, 21, 26, 154, 198
Hawking, Stephen, 12, 151
Heidegger, Martin, 51, 80, 142
Heisenberg, Werner, 82, 123, 155
Hellman, Geoffrey, 126

Helmholtz, Hermann von, 158
Hempel, Carl, 215
Heraclitus, 143
Herder, Johann, 169
Hertz, Heinrich, 217
Hespos, Simon, 16
Higginbotham, James, 32, 222, 238, 239
Hinzen, Wolfram, 10, 21, 23, 32, 35, 49, 51, 55, 60, 61, 62, 66, 67, 68, 79, 102, 129, 131, 132, 193, 198, 200, 205, 209, 214, 215, 218, 220, 222, 227, 229, 235, 237, 238, 240, 241, 243, 244, 246, 247, 248, 249, 252, 261
Hobbes, Thomas, 4, 226, 230
Holmes, Sherlock, 236
Homogeneity Hypothesis, 8
Hornstein, Norbert, 23, 204
Horwich, Paul, 24
Hovav, Malka, 242
Huineng, 177
Hull, David, 68
Humberstone, Lloyd, 211
Humboldt, Wilhelm von, 13, 25, 26, 33, 34, 86, 102, 135, 199, 225
Hume, David, 4, 12, 18, 26, 67, 81, 82, 85, 105, 116, 132, 136, 137, 141, 146, 167, 171, 210, 212, 218, 226, 228, 230
Humphrey, Nicholas, 171
Husserl, Edmund, 44, 145
Hut, Piet, 79
Huxley, Aldous, 6, 89, 90, 124, 133, 134, 136, 143, 148, 168, 191, 217
Huxley, Julian, 44
Huxley, Thomas, 53, 157
Huygens, Christiaan, 152, 164
Jackendoff, Ray, 17, 18, 22, 24, 34, 51, 52, 59, 64, 65, 92, 99, 116, 117, 128, 193, 194, 211, 215, 216, 220, 224, 226, 227, 229, 231, 233, 234, 235, 236, 238, 242, 243, 244, 250
Jackson, Frank, 47, 211

Jackson, J. Hughlings, 34
Jacob, François, 34, 80
Jacob, Margaret, 143
Jacob, Pierre, 97, 120, 145, 152, 195, 197, 211, 223
James, William, 7, 48, 156, 158, 247
Jaynes, Julian, 171
Jeannerod, Marc, 211
Jespersen, Otto, 16
Johnson, Mark, 19, 198, 215, 248
Johnson, Samuel, 25
Joyce, James, 77, 217
Jung, Carl, 172
Kaku, Michio, 152
Kandel, Eric, 160
Kant, Immanuel, 13, 96, 105, 155
Kaplan, David, 211
Karamazov, Ivan, 101
Katz, Jerold, 22
Kauffman, Stuart, 49
Kayne, Richard, 23
Keats, John, 138
Keleman, Deborah, 234
Keller, Helen, 36
Kellerman, Eric, 8
Kelvin, Lord William Thomson, 136
Kepler, Johannes, 143, 152
Kim, Jaegwon, 125, 153
Kinsbourne, Marcel, 161
Klein, Melanie, 176
Kleist, Heinrich von, 70
Koffka, Kurt, 226
Kohler, Ivo, 159
Köhler, Wolfgang, 226
Kolmogorov, Andrey, 80
Konner, Melvin, 171
Kripke, Saul, 120, 125, 201, 211, 212, 213, 229, 236, 241
Kropotkin, Peter, 65
Kuhl, Patricia, 37
Kuhn, Thomas, 149, 246
Kundera, Milan, 111
Laden, Anthony Simon, 95
Lakoff, George, 19, 20, 230
Landau, Barbara, 113
Langacker, Ronald, 227

Lange, Friedrich, 29, 82, 88, 96, 100, 111, 126, 128, 130, 131, 136, 137, 147, 205, 213
LaPorte, Joseph, 211
Laurie, Hugh, 25
Lawrence, D. H., 95, 142
Lawton, Graham, 159
Lear, King, 236
Leavis, Frank, 207
Lederer, Richard, 114
Leibniz, Gottfried, 4, 13, 93, 125, 132, 137, 154
Lenneberg, Eric, 45
Leonardo da Vinci, 115, 158
Lepore, Ernest, 193
Leuthardt, Eric, 92
Levin, Beth, 242
Levin, David M., 192
Levins, Richard, 52
Lewin, Kurt, 52
Lewontin, Richard, 15, 33, 52, 56
Libben, Gary, 8
Libet, Benjamin, 85, 87
Lightfoot, David, 97
Llinás, Rodolfo, 156, 175, 220, 221
Locke, John, 4, 21, 67, 78, 83, 100, 112, 129, 132, 136, 139, 146, 147, 149, 159, 166, 206, 216, 220, 222, 226, 230
Looney Tunes, 164
Lowe, E. J., 100, 142
Lucretius, 44, 84, 91, 136, 137
Lucy, John A., 21
Ludlow, Peter, 70, 205, 222
Lumsden, C. J., 102
Lycan, William G., 80, 145
MacLance, Saunders, 122
MacNamara, John, 231
Magritte, René, 124
Malt, Barbara, 203
Marcel, Gabriel, 165
Marr, David, 199, 218, 232
Marr, Edward, 200
Marsais, César Chesneau de, 196
Marx, Karl, 48
Maudsley, Henry, 228
Maugham, W. Somerset, 34

Mayr, Ernst, 53
McCorquodale, Marjorie, 134
McDowell, John, 210
McGilchrist, Iain, 29, 69, 70, 113, 118, 134, 141, 161, 162, 165, 168, 169, 170, 177, 235
McGilvray, James, 40, 52, 61, 102, 103, 119, 161, 192, 193, 194, 198, 204, 208, 219, 223, 225
McGinn, Colin, 25, 77, 89, 94, 121, 218
Mead, G. H., 31
Mello, Craig, 86
Melnyk, Andrew, 129, 201
Meltzoff, Andrew, 99
Merleau-Ponty, Maurice, 159, 166
Metzinger, Thomas, 88
Metz-Lutz, Marie-Nöelle, 98
Mikhail, John, 67
Miles, Lynden, 113
Mill, John Stuart, 14
Miller, James, 37
Milner, Marion, 168
Minimalism, 46
Monod, Jacques, 91, 101, 102
Montague, Richard, 216
Montaigne, Michel de, 14, 174
Moore, George Edward, 238
Morris, Charles, 22
Muir, John, 165
Nagel, Thomas, 146, 171
Napier, John, 28
Nathanaël, 29
Nelson, Katherine, 9
Neta, Ram, 214
Neumann, John von, 61
Neurath, Otto, 135
Newton, Isaac, 4, 46, 58, 83, 97, 121, 125, 126, 128, 129, 132, 135, 136, 137, 138, 139, 141, 142, 143, 146, 147, 148, 149, 156, 157, 163, 247
Nietzsche, Friedrich, 87, 160, 204
Núñez, Rafael, 122
O'Shaughnessy, Brian, 156
O'Toole, Peter, 214
Olschki, Leonardo, 53

Olson, Eric T., 166
Only Fools and Horses, 228
Owen, Richard, 49, 51
Parmenides, 138, 235
Partee, Barbara, 216
Pascal, Blaise, 250
Pasteur, Louis, 150
Pauli, Wolfgang, 247
Pauling, Linus, 127
Peat, F. David, 160
Piatelli-Palmarini, Massimo, 52
Pierce, Charles Sanders, 105
Pietroski, Paul, 24, 32, 57, 63, 192, 201, 210, 240, 244
Pinker, Steven, 48, 51, 59, 64, 65, 66, 68, 82, 112, 120, 130, 134, 137, 178, 203, 204, 219, 242, 243, 244, 245
Planck, Max, 216, 251
Plato, 6, 10, 70
Plutarch, 228, 229
Poeppel, David, 248
Poincaré, Henri, 47, 135
Poland, Jeffrey, 128, 141
Popper, Karl, 133
Priestley, Joseph, 4, 82, 116, 127, 133, 138, 139, 141, 196, 202
Pustejovsky, James, 193
Putnam, Hilary, 138, 198, 202, 210, 211, 213, 214, 229
Pylyshyn, Zeno, 23
Pynchon, Thomas, 136
Quine, W. V. O, 11, 31, 118, 125, 133, 238
Ramachandran, V. S., 153, 173
Ramsey, Frank, 246
Reclus, Élisée, 251
Reichenbach, Hans, 239
Reid, Thomas, 100, 207
Rey, Georges, 157
Rickles, Dean, 124
Rizzi, Luigi, 61
Rizzolatti, Giacomo, 173
Robson, David, 92, 118, 149, 158
Rolland, Romain, 168
Rorty, Richard, 114, 147
Rosch, Eleanor, 196
Rosenkrantz, Gary, 193

Rotenberg, V. S., 29
Rousseau, Jacques, 46, 47, 86, 157
Rovane, Carol, 200
Rovelli, Carlo, 125
Russell, Bertrand, 7, 11, 13, 32, 47, 54, 77, 80, 81, 91, 96, 103, 115, 130, 131, 138, 142, 145, 148, 155, 162, 176, 191, 213
Russell, Edward Stuart, 49
Sacks, Oliver, 69
Salinger, J. D., 101
Sapir, Edward, 28, 113, 191, 200, 205
Sartre, Jean-Paul, 51, 165
Sass, Louis, 169
Scheler, Max, 175
Schlesinger, Arthur M., 90
Schopenhauer, Arthur, 12, 81, 85, 152, 157
Schrödinger, Erwin, 123, 128
Searle, John, 22, 33, 87, 103, 129, 163, 203, 236
Segalowitz, Norman, 17
Sennert, Daniel, 50
Seung, Sebastian, 86
Severtzov, Nikolai, 49
Shakespeare, 89, 156
Shelley, Percy Bysshe, 88, 138
Shepard, Roger, 79
Ship of Thesues, 228
Shklovsky, Viktor, 169
Simpson, G. G., 50
Sinha, Pawan, 100
Sklar, Robert, 45
Smith, Adam, 86
Smith, Barry C., 164
Smith, Neil, 119
Smolin, Lee, 89
Snell, Bruno, 237
Soames, Scott, 211
Socrates, 82, 148, 170, 214, 230
Spelke, Elizabeth, 16, 30, 31, 138, 244
Spencer, Herbert, 48, 65
St. Hilaire, Geoffroy, 49, 53
Stalnaker, Robert, 211, 216
Stapel, Elizabeth, 55
Stenger, Victor, 151

Stich, Stephen, 138
Stockwell, Peter, 112
Stone, Tony, 34
Stotz, Karola, 214
Strawson, Galen, 12, 23, 79, 124, 140, 144, 149, 205, 208, 210, 217, 223, 229, 248
Strawson, Peter, 154, 191, 192, 195, 196, 197, 231
Swift, Graham, 90
Talmy, Leonard, 230
Tanner, Michael, 81
Tarski, Alfred, 218
Tennyson, Alfred, 98
The Deer Hunter, 194
Thompson, D'Arcy, 49, 55, 56, 78, 193
Tolstoy, Leo, 121
Tooby, John, 98
Toulmin, Stephen, 133
Turing, Alan, 45, 55, 56, 78, 85, 93, 116, 117
Uganda, 198
Underhill, James, 101, 199
Vaucanson, Jacques de, 116
Veblen, Thorstein, 166
Vedral, Vlatko, 123
Vignale, Giovanni, 142, 145
Voltaire, 157
Wajcman, Gérard, 4
Wallace, Alfred Russel, 50, 56
Warren, Josiah, 95
Wartofsky, Marx, 48, 81, 139, 141, 235
Waterloo, Battle of, 132
Weinberg, Steven, 146
Weismann, August, 48
Wernicke's area, 92
Wetheimer, Max, 226
Whitehead, Alfred North, 131
Whitman, Walt, 114
Whorf, Benjamin Lee, 113
Wilczek, Frank, 151
Williams, George, 66
Wilson, Edward O., 234
Wilson, Timothy D., 154, 166

Wittgenstein, Ludwig, 6, 21, 31, 102, 145, 148, 158, 169, 170, 177, 202, 205, 217, 223, 225
Wittreich, Warren J., 159
Woozley, A. D., 100
Xenophanes, 102

Yankama, Beracah, 24, 32, 57, 63
Yolton, John, 209
Zamyatin, Yevgeny, 77, 121
Zeki, Semir, 160
Ziman, John, 131

Printed in Great Britain
by Amazon.co.uk, Ltd.,
Marston Gate.